Chemical Infrastructure Protection and Homeland Security

FRANK R. SPELLMAN AND
REVONNA M. BIEBER

GOVERNMENT INSTITUTES
An imprint of
THE SCARECROW PRESS, INC.
Lanham • Toronto • Plymouth, UK
2009

Published in the United States of America
by Government Institutes, an imprint of The Scarecrow Press, Inc.
A wholly owned subsidiary of
The Rowman & Littlefield Publishing Group, Inc.
4501 Forbes Boulevard, Suite 200
Lanham, Maryland 20706
http://www.govinstpress.com/

Estover Road
Plymouth PL6 7PY
United Kingdom

The reader should not rely on this publication to address specific questions that apply to a particular set of facts. The author and the publisher make no representation or warranty, express or implied, as to the completeness, correctness, or utility of the information in this publication. In addition, the author and the publisher assume no liability of any kind whatsoever resulting from the use of or reliance upon the contents of this book.

British Library Cataloguing in Publication Information Available

Library of Congress Cataloging-in-Publication Data

Spellman, Frank R.
Chemical infrastructure protection and homeland security / Frank R. Spellman and Revonna M. Bieber.
p. cm.
Includes bibliographical references and index.
ISBN 978-0-86587-182-3 (pbk. : alk. paper) — ISBN 978-1-59191-945-2 (electronic)
1. Chemical plants—Risk assessment—United States. 2. Chemical industry—Security measures—United States. 3. Chemical industry—Defense measures—United States. 4. Civil defense—United States. 5. Terrorism—United States—Prevention. I. Bieber, Revonna M., 1976– II. Title.
TP155.5.S63 2009
363.325'96600973—dc22 2009002472

∞™ The paper used in this publication meets the minimum requirements of American National Standard for Information Sciences—Permanence of Paper for Printed Library Materials, ANSI/NISO Z39.48-1992.
Manufactured in the United States of America.

For JoAnn Garnett-Chapman,
the ultimate professional

Contents

Preface

The third of a new Government Institutes series on critical infrastructure and homeland security, *Chemical Infrastructure Protection and Homeland Security* is a reference source that serves chemical manufacturing/processing/production businesses and managers who want quick answers to complicated questions, to help employers and employees handle the security threats they must be prepared to meet on a daily basis. In the post–September 11 world, the possibility of chemical terrorism—the malicious use of weapons to cause devastating damage to the chemical industrial sector, along with its cascading effects—is very real.

Prior to 9/11, an assessment of chemical plant site security by the Agency for Toxic Substances and Disease Registry (ATSDR) was considered by many to be the most comprehensive analysis that was publicly available. ATSDR researchers reviewed national statistics on domestic terrorism compiled by the FBI in 1995 and interviewed security staff from facilities and potential targets in one community with numerous chemical plants. Reviewers found the following concerns:

- Security at chemical plants ranged from fair to very poor.
- Chemical plant security managers were very pessimistic about their ability to deter sabotage by employees, yet none of them had implemented simple background checks for key employees such as chemical process operators.
- None of the corporate security staff had been trained to identify combinations of common chemicals at their facilities that could be used as improvised explosives and incendiaries.

The full ATSDR report, which was never made public, noted that

> among the "soft targets" that the ATSDR identified as potential terrorist sites were chemical manufacturing plants (chlorine, peroxides, other industrial gases, plastics, and pesticides); compressed gases in tanks, pipelines, and pumping stations; and pesticide manufacturing and supply distributors.

This book (and all forthcoming volumes of the critical infrastructure series) was written as a result of 9/11 to address these concerns. *Chemical Infrastructure Protection and Homeland Security*, in particular, was fashioned in response to the critical needs of chemical production managers, chemical import managers, chemical industry infrastructure engineers, design engineers, process managers at any level of chemical production, students—and anyone with a general interest in the security of our chemical supply systems. It is important to point that our chemical industry (as is the case with the other fourteen critical infrastructures) cannot be made immune to all possible intrusions or attacks; thus, it takes a concerted, well-thought-out effort to incorporate security upgrades in the retrofitting of existing systems and careful security planning for all new chemical processing sites. These upgrades or design features need to address issues of monitoring, response, critical infrastructure redundancy, and recovery to minimize risk to the facility/infrastructure.

Chemical Infrastructure Protection and Homeland Security presents common-sense methodologies in a straightforward, almost blunt manner. Why so blunt? At this particular time, when dealing with security of workers, family members, citizens, and society in general—actually, with our very way of life—politically correct presentations on security might be the norm, might be expected, and might be demanded. Frankly, our view is that there is nothing normal or subtle about killing thousands of innocent people; mass murders certainly should not be expected; and the right to live in a free and safe environment is a reasonable demand.

This text is accessible to those who have no experience with the chemical industry and/or homeland security. If you work through the text systematically, you will gain an understanding of the challenge of domestic preparedness—that is, an immediate need for a heightened state of awareness of the present threat facing the chemical industrial sector as a potential terrorist target. Moreover, you will gain knowledge of security principles and measures that can be implemented, adding a critical component not only to your professional knowledge but also giving you the tools needed to combat terrorism in the homeland—our homeland.

Frank R. Spellman
Revonna M. Bieber
Norfolk, Virginia

REFERENCES AND RECOMMENDED READING

Agency for Toxic Substances and Disease Registry (ATSDR). Industrial chemicals and terrorism: Human health threat analysis, mitigation, and prevention. www.mipt.org/pdf/industrialchemicalsandterrorism.pdf (accessed May 17, 2008).

1

Introduction

> Facilities handling large amounts of potentially hazardous chemicals (i.e., chemical facilities) might be of interest to terrorists, either as targets for direct attacks meant to release chemicals into the community or as a source of chemicals for use elsewhere. Because few terrorist attacks have been attempted against chemical facilities in the United States, the risk of death and injury in the near future is estimated to be low, relative to the likelihood of accidents at such facilities or attacks on other targets using conventional weapons. For any individual facility, the risk is very small, but the risks may be increasing—with potentially severe consequences for human health and the environment. Available evidence indicates that many chemical facilities may lack adequate safeguards. (Shea 2008)

WE CALL IT 9/11

Before the first World Trade Center bombing in 1993 and the 1995 Oklahoma City bombing, emergency incidents were primarily or generally thought to be caused by natural or accidental events. Examples of natural phenomena (events) include wildfires, flash floods, earthquakes, active volcanoes, droughts, and storms. These natural events are not entirely predictable, and they cannot yet be controlled or prevented (Meyer 2004).

Accidental incidents involving new types of emergencies began to surface in the 1940s. They were linked with the behavior of certain chemical products collectively called hazardous chemicals (or hazardous materials) whenever they are misused or involved in unintended mishaps and fires.

Where there are people, there are hazardous chemicals—from the busiest metropolis to the most remote community. In their myriad forms, hazardous chemicals are at the heart of our technology-based society. Chemical products can make our lives longer, healthier, and more productive, prosperous, and comfortable. Chemical products are essential to the U.S. economy and standard of living; they allow us to live or experience the trappings of the "good life." The chemical industry manufactures products that are fundamental elements of other economic sectors. For example, it produces fertilizer for agriculture, chlorine for water purification, and polymers that create plastics from petroleum for innumerable household and industrial products. The most common of these include plastics, leather, paper, rubber, paints, textiles, pesticides, solvents, detergents, fuels, medicines, fertilizers, building materials, electronics, sporting equipment, and automobiles. In addition, it is important to point out that an enormous amount of the chemical industry's products go to health care. To eliminate hazardous chemicals from our society would not only be impractical, it would be undesirable (Meyer 2004). Instead, since it is impractical and undesirable to eliminate hazardous chemicals from our society, we must find ways to work and live safely with these hazardous chemicals and products.

A third type of emergency incident became apparent to the public on September 11, 2001. On "9/11," as it is now called, terrorists struck at the heart of America, on American soil, in a way that is unforgettable to all of us. Airplanes filled with people and fuel were turned into guided missiles of death and destruction. Keep in mind that 9/11 did not change the nature of hazardous chemicals; they have always been hazardous. Nor did the events of that horrendous day change the way in which we respond to such incidents. What the 9/11 terrorists provided us with was a general awareness of a type of extremely deadly venom of hate delivered by groups or individuals that is designed to kill massive numbers of people, cause substantial property damage, and affect economic stability. Most importantly, for the public and emergency responders, the events of 9/11 brought to the forefront knowledge of the presence of biological and chemical warfare agents within the communities in which we live; we have learned to anticipate their presence. In regard to terrorism and terrorist acts upon our homeland, we have learned that just about anything is possible. Thus, security of the homeland is vital to all of us.

Governor Tom Ridge, a U.S. political figure who served as a member of the U.S. House of Representatives (1983–1995), governor of Pennsylvania (1995–2001), assistant to the president for homeland security (2001–2003), and the first U.S. secretary of homeland security (2003–2005), got it right when he stated, "You may say Homeland Security is a Y2K problem that doesn't end Jan. 1 of any given year" (Henry 2002). Homeland security is an ongoing problem that must be dealt with 24/7. Simply, there is no magic on-off switch that we can use to turn off the threat of terrorism in the United States or elsewhere.

The threat to our security is not only ongoing but also universal, including potential and real threats from within—from our own citizens (homegrown terrorism). Consider the American Timothy McVeigh, for example, who blew up the Alfred P. Murrah Federal Building in Oklahoma City in 1995, killing almost 200 people, including several children. McVeigh, who bombed the building in revenge for the FBI's Waco,

Texas, raid, thought the army (he was a decorated U.S. Army veteran) had implanted a chip in his butt to track his movements, according to reports.

It is interesting to note that McVeigh, who was no doubt suffering from some type of severe disturbance, acted primarily alone. Actually, McVeigh is the exception that proves the rule—most terrorist acts on America are planned by a group beforehand. However, this is not always the case. For example, consider the case study that follows—an incident that occurred in 1992, before McVeigh, before the first attack on the World Trade Center, before 9/11, and prior to the anthrax attacks.

Sidebar 1.1. When Sugar Leaves a Bitter Taste

My name is W. W. Williams . . . just call me Willy, thank you very much. Normally, I would not be having this conversation with you . . . no, sir, I wouldn't. I've never been about talking or jiving or other such dribble. No, sir. Talk is cheap and action is expensive . . . my kind of action anyway.

Well, let's cut to the chase. Not that I have anything else more pressing to do. No, sir. It's just that I am not used to being out of my psychic ether mode . . . from which I perpetually surfed . . . until I got caught, that is.

Now, at the present time, I am disgustingly sober and in total control of my wits and sanity. But I am not free. No, sir. I sit here on my cement bed in Supermax, Colorado, . . . the ultimate prison for the world's worst prisoners. Yes, I am one of the worst . . . really bad, but sickeningly sober.

Speaking of soberness, I am one of the Pied Piper's (Hunter S. Thompson) followers. It was Thompson who described the good life . . . the psychic Ethernet at its best! In the fog of soberness, I have memorized Thompson's famous description of his driving orgy through Las Vegas.

> In the trunk, we "stow two bags of grass, seventy-five pellets of mescaline, five sheets of high-powered blotter acid, a salt shaker half full of cocaine, and a whole galaxy of multicolored uppers, downers, screamers, laughers, . . . a quart of tequila, a quart of rum, a case of Budweiser, a pint of raw ether and two dozen amyls."

Doesn't it just torque your jaws that some of us know how to live . . . and the rest just exist. God, what a man. . . . Thompson is my all-time hero . . . for sure.

Sigh! I guess we should get back to the topic and conversation at hand. I was given this chance to state my case . . . to explain why I blew up the sugar production factory and killed all those common sufferers out there in la-la land. So, let's get to it. Let me tell you my story. It ain't pretty in some folks' way of thinking, I suppose. But that . . . others' way of thinking . . . has never bothered me too much. . . . No, sir. I do it and did it my way. And I would be happy forever sitting on this cement bed speaking to you all if they would just let me have my ether. Oh well, what the hell . . . can't have it all . . . not anymore, anyway.

(*continued*)

I worked at the Admiral Sugar Processing Plant for eleven years before I blew the place to smithereens. No, sir, I did not want to blow the place up . . . but I have to tell you it was as easy as rolling a joint . . . if you get my drift?

I did not want to blow the place up because I had a good job there making decent money . . . more than enough to keep me surfing the psychic ether constantly. Hell, what else can one ask for? Well, as it turned out, we can ask for more. I found this out when, after ten years of dedicated, loyal, and exceptional service, they promoted someone over me for the superintendent's position.

Being passed over was a slap in the face. Being passed over by a woman . . . an Asian woman to boot . . . is the ultimate slap in the face . . . the final emasculation of a red-blooded all-American male . . . who just happened to be one of Admiral's top performers. Well, that was certainly the case any time when I was surfing the psychic ether.

I put up with working for Ying Yang (that's what we called her because she said she had a twin sister and she was Asian—the name made sense to us but she hated it but had no choice . . . the poor devil, ha) for one year. Then it came to me one night in one of my clear-headed stupors that I needed to blow that place up. I did.

No, sir, as I stated earlier it was no big thing . . . it was easy. The fact is most people have no idea, no clue of what a sugar plant is all about. The common folk out there just do not understand that a sugar plant is full of sugar dust . . . bad stuff. Anyway, sugar dust is so combustible that static electricity, sparks from metal tools, or a cigarette can ignite explosions. During a safety training session that the female safety manager (we called her the safety wench) conducted there (she died in the explosion; they never found her remains; no big loss in my opinion) explained to us that more than three hundred dust explosions have killed more than a hundred workers in grain silos, sugar plants, and food processing plants in the past three decades. Most are preventable by removing fine-grain dust as it builds up. The safety wench explained what she called the dust explosion pentagon to us. We all had the following facts drilled into our heads:

> In addition to the familiar fire triangle of oxygen, heat, and fuel (the dust), dispersion of dust particles in sufficient quantity and concentration can cause rapid combustion known as a deflagration. If the event is confined by an enclosure such as a building, room, vessel, or process equipment, the resulting pressure rise may cause an explosion. These five factors (oxygen, heat, fuel, dispersion, and confinement) make up the sides of the dust explosion pentagon. If one element of the pentagon is missing, an explosion cannot occur.

Normally, at Admiral Sugar Processing Plant the five factors of the dust explosion pentagon are not present at the same time; thus, as mentioned above, a deflagration is impossible. However, ever since my having been passed over for advancement and the promotion of Ying Yang, all of the men at the plant (including me, of course) basically ignored her. Yes, she was the boss and she walked the plant barking orders to all of us but we just continued our work and ignored her. She knew we were ignoring her but she was smart enough not to elicit confrontation with any of us men. She just sort of made her walk around (made an appearance so it looked like she was doing her job) and then went back into her office, closed the door, and hibernated (or whatever it is women do when they are ignored by all) until it was time to make the final walk-around and then head for home.

During Ying Yang's tenure, Admiral Sugar Processing Plant continued to process sugar and to meet production goals. No, sir, production was not the problem . . . we all valued our jobs and paychecks so even though we disliked Ying Yang we still performed to keep up production . . . and appearances—at least in my case. I had already decided that I would get even with upper management for not selecting me as plant superintendent.

I am the patient type. No, sir, I do not rush into anything . . . well, there are some things, I guess. Anyway, because of Ying Yang's dysfunctional management style (have you ever noticed how many dysfunctional managers there are?—gee, they are everywhere . . . too many to count). . .

Anyway, because Ying Yang overlooked the workers' refusal to keep housekeeping up to where it always had been . . . where the safety wench insisted that it be, there were certain obscure, hidden compartments (not totally confined spaces like manholes per se . . . more like snug, cramped cubbyholes) in our main factory building where sugar dust just kept accumulating to the point where it was six to eight inches deep in spots. And while my master plan was in the works the safety wench was off to the other coast getting some advanced OSHA training.

So, after two weeks I had ensured that two legs of the dust explosion pentagon were in place, the fuel (sugar) and the confined areas (cubbyholes). The third leg, oxygen, also was not a problem because even though all of the cubbyholes were located in the east side of the plant where few workers worked, ubiquitous oxygen was in fresh supply throughout the plant site, thanks to a generously designed dilution ventilation system.

I did, however, identify a problem with the fourth leg of the dust explosion pentagon, the dispersion of the fine sugar dust. The fact was dispersion of fine sugar dust (fuel) was not allowed or present in any of the cubbyhole areas . . . even though there was a huge source of sugar available. I noticed that all of the cubbyholes were, in regard to air flow, dead areas . . . and this was by engineering design, of course. The dead space was deliberately designed so as to prevent the occurrence of deflagrations like I was attempting to create.

Thus to ensure dispersion and airborne suspension of the fine sugar dust (fuel), I had to investigate each cubbyhole configuration and figure out how to disperse the fuel . . . make it flow (drift) to fill each nook and cranny of the two football field–sized east sections of the plant building.

I found the answer. All along the east side of the plant building the entire wall surface had been painted flat black. As I walked along the wall, I noticed that

(*continued*)

about every twelve feet or so the flat black smoothness of the wall surface was interrupted with rectangular-shaped windows. On closer inspection, I noticed that these windows (I had counted sixty-six of them) were exactly four feet above floor level and separated by eight feet. They looked to be about five feet in height. Each metal window frame and its accompanying wire-mesh glass had been painted flat black like the rest of the wall. I also noticed that each window operating handle, used to jack open each window, had been removed. In addition, screw holes with screws, washers, and nuts had been inserted (and painted over) into each window at various locations along the window frame to ensure that the windows remained closed—as configured, they could not be opened.

I spent the next two weeks while on my breaks, during lunch, and whenever I could sneak away from my work area for a few minutes removing the locking screws from several window frames. Never mind how many . . . gee, who the heck was counting?

After each makeshift locking device was removed from each window, I used a small, thin pry bar to wedge open each window about an eighth of an inch . . . just enough to let outdoor air seep through the window frames to help stir up the dust and keep it dispersed throughout the east end but not open enough so that anyone would notice (that was my intent, anyway).

Over the next few days I noticed that the sugar dust was dispersed better and no one else seemed to notice. Each shift custodial staff that cleaned during each of the three shifts always ignored the east end of the building because they were lazy and had more pressing cleanup concerns in the actual production areas. Thus, thanks to my efforts, and as the days went by, the sugar dust accumulated and dispersed throughout the east end . . . quite nicely, actually.

At this stage of my master plan, the only leg of the dust explosion pentagon missing was the spark or heat source. I thought about this leg for some time . . . a few days or so. Many possibilities ran though my mind: a stick of dynamite, TNT, Bic lighter, cigarette wrapped in matches, and others. I found an almost endless number of possibilities on the Internet. It is unbelievable how helpful cyberspace is in devising a master plan of death and destruction. I was quite impressed.

Anyway, I finally came up with the heat source. From outside the building's east end I had noticed a couple of places running along the outside wall where 115-volt all-weather type electrical sockets had been installed. All the sockets were dead or deenergized . . . no electricity. I discovered this when I used a small lamp that I plugged into the sockets during a couple of my breaks. Every socket was without electrical power . . . apparently someone had deenergized them many moons ago. Anyway, I looked around outside and found a cement block house that served as the electrical utility room. Inside the dinky, dim-lighted space were walls covered with electrical power panels of various sizes. Each of the power panels contained circuit breakers for specific plant equipment, sections of overhead lighting, and assorted electrical outlets within the plant building. It did not take long to find the power panel I was looking for. When I opened the panel door, I noticed that all six of the circuit breakers within had their circuit breaker handles removed; the resultant handle holes were covered over with small pieces of fiberboard.

A few days later, I got the chance to search for and find a spare circuit breaker with on-off switch handle attached. I removed the spare breaker and eventually installed it into one of the empty breaker sockets in the power panel for the outdoor 115-volt outlets.

The next problem I had was to determine which of the electrical outlets I had actually re-energized. Using the portable lamp again, I went from socket to socket until I found three outlets that were now energized. I now had a source of electrical power to provide the heat needed to cause the deflagration.

I found an old sixty-foot extension cord in my garage. I cut the female plug off one end and twisted the bare copper ends of the white and black wires together. This formed the short circuit I wanted. Simply, I knew that when the cord was energized, the short-circuited wires would produce an instantaneous heat source, setting off the deflagration.

I ran the short-circuited end through one of the windows I had wedged open and placed the twisted-together copper wire section on top of a four-inch pile of sugar. The plug-in end ran along the ground along the outside wall to the first energized outlet; I did not plug in the extension cord. I turned off the circuit breaker and left it off.

The stage was finally set . . . all legs of the deadly dust explosion pentagon were now in place.

I waited a couple of days until April 20, which I had chosen as my get-even-with-the-world-day. Historically, big things seem to happen on this particular date—the birth of Hitler, the Oklahoma City bombing, the Columbine High School massacre, and the Waco showdown—so I was determined to continue the trend.

About 10:30 in the morning, April 20th, I wandered outside the plant and over to where I had stashed the plug-in. Before plugging in the extension cord I checked the circuit breaker in the block house to ensure it was in the OFF position. Then I plugged in the detonation cord and went back into the block house and shut the steel door tight . . . I wanted to make sure I did not blow up in the explosion.

All I remember is turning on the circuit breaker and hearing what I thought was a distant rumble and a slight shaking of everything and then total darkness . . . nothingness.

Two days later I woke up in the hospital bed with all kinds of tubes and machines plugged into me and a couple of armed guards at the door. I seemed to be in one piece but I was not exactly sure at that moment.

After a few days I had my situation pretty well figured out. I had miscalculated the effect of the explosion and it had blown everything to hell . . . and the block house

(*continued*)

that I thought I was safe inside was severely damaged also . . . but it had saved my life because though I suffered a few broken bones and some minor internal injuries, I was just fine. However, the authorities pretty much put two and two together on how the explosion had occurred and who had thrown the switch . . . that would be me.

So, here I sit two years later and after one of the longest, most ridiculous trials in criminal history. Everyone was out to get me. I had killed eleven and injured forty-two. Some of the injuries were so severe and disfiguring that many of the women who testified in court looked like they were kissing cousins of Freddie Krueger, one of my all-time heroes.

Sitting here now and basking in the glory of what I accomplished, only two things bother me at the moment: one, I got caught; two, during the trial and all the publicity that followed, they tried to classify me as an American terrorist. Terrorist? No way. I am a mass killer who wishes he had killed more; I am not a terrorist. I really don't think the press, law authorities, or the people know what a terrorist really is. Maybe someone ought to write a book on the subject to inform everyone what a terrorist really is. One thing is certain; no, sir, I am not a terrorist.

The incident related in sidebar 1.1 points to the purpose of this text—namely, to emphasize the importance of chemical industry production facilities (yes, sugar is a chemical, an organic chemical; sugar production qualifies as a chemical production process) being ready for any contingency—"any" being the key word.

While it is obvious that the statement above about sugar being a chemical is true, it is generally not so obvious to many environmental practitioners (those not familiar with dust explosion hazards) that the two major regulations/standards that provide guidelines on protecting against the hazards of hazardous chemicals/materials (OSHA's Process Safety Management Standard and USEPA's Risk Management Program) do not pertain to sugar dust. Sugar dust, even though highly volatile, is not a "listed" hazardous chemical/material. This is also the case with other combustible dusts that may cause a deflagration, other fires, or an explosion. In addition to organic sugar dust, these dusts include, but are not limited to:

- metal dust such as aluminum and magnesium
- wood dust
- coal and other carbon dusts
- biosolids
- other organic dust such as paper, soap, and dried blood
- plastic dust and additives
- certain textile materials

Because the Process Safety Management Standard and Risk Management Program do not address the hazards related to the dusts listed above, other regulatory means of ensuring safe and secure conditions had to be devised for specific industries. In light of this and because a combustible hazard study conducted by the U.S. Chemical Safety and Hazard Investigation Board (CSB) found that nearly 280 dust fires and explosions have occurred in U.S. industrial facilities over the past twenty-five years, resulting in 119 fatalities and over 700 injuries, OSHA initiated a Combustible Dust National Emphasis Program (NEP). OSHA issued this NEP directive under number CPL 03-00-006 (OSHA 2007).

According to OSHA (2007), the purpose of this NEP is to inspect facilities that generate or handle combustible dusts that pose a deflagration or other fire hazard when suspended in air or some other oxidizing medium over a range of concentrations, regardless of particle size or shape; deflagrations can lead to explosions. Combustible dusts are often either organic or metal dusts that are finely ground into very small particles, fibers, fines, chips, chunks, flakes, or a small mixture of these.

In situations where the facility being inspected is not a grain-handling facility, the lab results indicate that the dust is combustible, and the combustible dust accumulations not contained within dust control systems or other containers, such as storage bins, are extensive enough to pose a deflagration, explosion, or other fire hazard, then citations under 29 CFR 1910.22 (housekeeping) or, where appropriate, 29 CFR 1910.176 (c) (housekeeping in storage areas) may generally be issued. Combustible dusts found in grain-handling facilities are covered by 29 CFR 1910.272.

In its NEP, OSHA points out that for workplaces not covered by 1910.272, but where combustible dust hazards exist within dust control systems or other containers, citations under section 5(a)(1) of the OSH Act (the General Duty Clause) may generally be issued for deflagration (fire) or explosion hazards. National Fire Protection Association (NFPA) standards should be consulted to obtain evidence of hazard recognition and feasible abatement methods. Other standards are applicable to the combustible dust hazard. For example, if the workplace has a Class II location, then citations under 29 CFR 1910.307 may be issued to those employers having electrical equipment not meeting the standard's requirements.

LISTED HAZARDOUS CHEMICALS/MATERIALS

In 1990–1991, American industries involved in hazardous chemicals/materials production, shipping, storing, and usage were scrutinized from both a safety and security viewpoint, to the point where OSHA's Process Safety Management Standard (PSM) 29 CFR 1910.119 was generated to address these concerns. PSM was promulgated in 1992, to be fully complied with by May of 1997.

The PSM Standard lists several hazardous chemicals/materials affected by the regulation, requiring full compliance. OSHA's primary purpose (a lofty and important goal—some would say an impossible goal) is to protect the worker from accidents and illnesses in the workplace. PSM is simply an extension of this effort. It should be pointed out, however, that OSHA standards, like PSM, are designed, almost exclusively, to protect workers within workplace fence lines. Herein lies the problem with PSM. Hazardous materials spills, accidental or intentional, often cross workplace fence lines into surrounding neighborhoods, affecting all three environmental media: air, water, and soil. Thus, while USEPA saw the benefit of PSM and applauded OSHA's efforts in this critical safety area, it also saw this critical drawback or shortcoming. Simply, hazardous chemicals/materials incidents that begin in the workplace are likely to spread their lethal effects beyond the fence line. This is the primary reason that USEPA "borrowed" many of the tenets of PSM and used that material to develop its own regulation, Risk Management Planning (RMP), 40 CFR Part 68. Both of these regulations, PSM and RMP, are designed to prevent and/or detail the proper mitigation procedures to be used during hazardous chemical/material incidents. The key point to remember is that PSM/RMP combined not only function to set requirements for *on-site* emergency response planning and training (to protect on-site personnel), but it also points out the need for emergency response planning for any *off-site* consequences (to protect the public).

What does all this information about PSM/RMP have to do with protecting chemical industry infrastructure? Good question. For those already familiar with the tenets of these important regulations, the answer is obvious. That is, in combination, these regulations set the basis, became the vital foundation for subsequent directives related to post-9/11 Homeland Security and protection of vital infrastructure, including chemical industry infrastructure, from terrorism—homegrown or otherwise.

To gain a better sense of the positive impact that compliance with PSM and RMP can have on facilities that use or produce covered hazardous materials in their processes (including chemical industry facilities), consider the previous disaster scenario from a different point of view. That is, the preceding sugar production plant incident, arguably a clear act of homegrown terrorism, could have developed in a much different manner if the elements and procedures required by both the 1991–1997 PSM and RMP or OSHA's 2007 NEP on Combustible Dust had been in place and had been followed. This is not to say that in 1991 the event could have been prevented. It is doubtful that it could have been prevented, because of our pre-9/11 mentality of that time. On the other hand, if PSM/RMP, the 2007 Combustible Dust NEP, and current Homeland Security Guidelines in effect today had been in effect in 1991, the event might have been prevented—nipped in the bud, guarded against.

In regard to listed hazardous chemicals/materials only, to prevent accidental releases and acts of terrorism (homegrown or otherwise) with these chemicals/materials tenets of PSM/RMP and recommendations provided by the Department of Homeland Security, many such incidents have a chance of being avoided or properly mitigated upon their initiation or actual occurrence. The obvious question is: Why and how?

In the first place, PSM and RMP require that all responsible parties survey their industrial complexes where covered chemical processes are employed and to closely scrutinize these processes to determine if any of the 130+ Highly Hazardous Chemicals listed in OSHA's PSM and/or if any of the 140+ Extremely Hazardous Substances listed in USEPA's RMP are stored, handled, used, or produced on-site (e.g., off-gases, etc.).

In many of the chemical catastrophes of the past, compliance with PSM and RMP would have ensured that the affected processes/plants had been thoroughly surveyed for processes using or producing highly hazardous chemicals and extremely hazardous substances. This survey would have determined, of course, that the sites used covered chemicals in their process. Moreover, the survey would have noted the normal quantity of listed chemicals stored on the plant site at any given time. This survey would also have made note if any quantities of listed chemicals stored on the plant site exceeded the PSM and RMP threshold quantities (TQ).

Along with a complete survey of covered chemicals, under PSM and RMP the affected site would be required to complete the following actions:

1. Plant management would have been required to perform a process hazard analysis (PHA) for the entire listed chemical process.
2. If the plant used or produced listed chemicals at levels greater than the TQ level listed in PSM and RMP regulations, plant management would be required to comply with all regulatory requirements as related in the regulations.
3. One of the important elements of PSM and RMP regulations that the plant would be required to comply with is on-site and off-site consequence analysis or modeling to assess potential on-site/off-site exposures.
4. Another PSM and RMP element is the requirement for emergency response planning. Emergency response planning is required for both on-site and off-site (public areas).
5. Plant operator training in HazMat Response is also required by PSM and RMP. The plant operators who respond to a fire or spill of any listed chemical must be trained on how to properly handle such incidents. This training element is absolutely critical in any emergency response incident.

Along with the constant threat posed by fire, there are also chemical disaster situations that can occur due to human error in system operation and/or a malfunction in system equipment. In addition, there are other emergency situations that must also be considered and planned for, including floods, hurricanes, earthquakes, tornadoes, volcanic eruption, snow/ice storms, avalanches, explosions, truck accidents, train derailments, airplane crashes, building collapses, bomb threats, riots, and vandalism.

Post-9/11, we need to add terrorism to the preceding long list of emergency situations. In regard to the Admiral Sugar Processing Plant incident, we need to realize that long gone are those days when railroad safety meant avoiding derailments and accidents. Today we must consider and dodge terrorist attacks from Molotov cocktails and improvised explosive devices (IEDs), as well as from armor-piercing bullets—and any other weaponry the nut cases can get their hands on.

Again, it is also important to remember that chemical emergency situations can easily reach beyond the boundaries of any industrial plant. This is to be expected, especially in this age of population explosion with its characteristic urban sprawl. It is not unusual to find, for example, a chemical industrial plant site or other industrial plant that originally was isolated from city dwellers but later became surrounded on all sides by neighbors. The point is that when a chemical spill or chemical disaster occurs in an isolated area there may be no cause for general alarm; however, when such a deliberate disaster occurs in the plant site as described in the sugar plant incident, it should be clear that the purpose of PSM, RMP, the Patriot Act, Homeland Security directives, OSHA's Combustible Dust NEP, and other safety/security factors is far-reaching—and absolutely critical to the survival of a free society.

Keep in mind that many define "free society" in different ways. This is important to us in this text because in our view in order to be truly safe from the terrorist threat (if that is possible, and we have our doubts) we must give up certain freedoms and accept closer scrutiny and vigilance in regard to actions that we normally assume to be no one's business but our own. There are differing opinions on this topic and we fully accept that. Again, we each define freedom in our own way—and this is our right as free people.

In regard to various definitions, to clearly understand the purview of this text requires the precise definition of some words that might be used more loosely in common conversation. The words presented here may differ from those used in the U.S. Department of Homeland Security National Infrastructure Protection Plan:

- *Vulnerability* is the manifestation of the inherent states of the system (e.g., physical, technical, organizational, social, cultural) that can be exploited by an adversary to adversely affect (cause harm or damage to) that system.
- *Intent* is the desire or motivation of an adversary to attack a target and cause adverse effects.
- *Capability* is the ability and capacity to attack and cause adverse effects.

- *Threat* is the *intent* and *capability* to adversely affect (cause harm or damage to) the system by adversely changing its states.
- *Risk* is the result of a threat with adverse effects to a vulnerable system (Haimes 2004, 699).

WHAT IS TERRORISM?

Since 9/11, we have heard it said by many of our students (and others) that there is controversy about the definition of the politically charged word *terrorism.* Actually, in our opinion, when you get right down to it, the controversy has more to do with the definition than (in most cases) with politics. Terrorism, like pollution, is a judgment call. For example, if two neighbors live next door to an air-polluting facility, one neighbor who has no personal connection with the polluting plant is likely to label the plant's output as pollution. However, the other neighbor who is an employee of the plant may consider the plant's pollution as dollar bills—dollars that are his livelihood. Thus, what we are saying here is that along with defining pollution, defining terrorism may be a judgment call, especially in the view of the terrorists.

To make our point on the countless differing views in defining terrorism, consider, for example, that if we were to ask a hundred different individuals to define terrorism, we would likely receive a hundred different definitions. As a case in point, consider the following: If we were to ask a hundred different individuals to describe the actions of Willy in the Admiral Sugar Processing Plant incident, how would they describe him and his actions? You might be surprised—we were. In 2000 and again in 2007, pre- and post-9/11, after reading Willy's sugar dust deflagration incident, a hundred randomly selected Old Dominion University Environmental Health juniors and seniors ("Generation Y" students ranging in age from twenty to forty-six years old) were asked to reply to a nonscientific survey questionnaire. Note that it is common practice to survey our students on pressing environmental and security issues of the day. The two questions and the students' responses to this unscientific survey are listed in table 1.1.

From the Old Dominion University survey it is clear that the students' perceptions of Willy's actions in the Admiral Sugar Processing Plant incident shifted dramatically from pre-9/11 to post-9/11. For example, when asked to select the best pre-9/11 descriptor to describe Willy, "crazy" and "insane" ranked high; however, after 9/11, the students' perception shifted away from "crazy" and "insane" to "terrorist." Likewise, the students' pre-9/11 responses on Willy's actions ranked high in "madness"; however, his actions post-9/11 overwhelmingly were described as "terrorism".

It is interesting to note that the 2000 year group reported prior to the September/October 2001 anthrax attacks and prior to 9/11; however, the student group's responses were provided after such events as the World Trade Center attack of 1993 and Timothy McVeigh's 1995 mass murder of the occupants of the Federal Building in

Table 1.1. Student responses, pre- and post-9/11

	Number of Responses	
*Student Response Descriptors**	*Pre-9/11 (2000)*	*Post-9/11 (2007)*
Question 1: In your opinion, Willy was		
crazy	30	2
a disgruntled former employee	10	0
insane	14	5
misguided	1	0
a cold-blooded murderer	13	15
a misfit	1	0
deranged	6	4
a lunatic	5	1
a bully	18	5
a terrorist	2	68
not sure	0	0
Totals	100	100
Question 2: In your opinion, Willy's actions are best described as		
madness	55	1
frustration	9	2
desperation	4	0
dysfunctional thinking	2	0
legitimate concern	1	0
threatening	6	1
terrorism	5	84
workplace violence	0	0
not sure	18	12
Totals	100	100

* Student response descriptors were provided to the students by the instructors.

Oklahoma City. This may help to explain why the year 2000 students were somewhat reluctant to describe Willy's actions as terrorism and/or to label him as a terrorist.

Terrorism by Any Other Name Is . . .

From the preceding discussion we might want to buy into the argument that terrorism is relative, a personal judgment. But is it really relative? Is it a personal judgment? What is terrorism? Take your choice. Seemingly, there is an endless list of definitions and no universally accepted definition of terrorism. However, since terrorism is the main theme of this text, let's review a few of these definitions.

Standard Definitions of Terrorism

After reviewing several dictionaries, we have found this fairly standard definition of terrorism:

> The unlawful use or threatened use of force or violence by a person or an organized group against people or property with the intention of intimidating or coercing societies or governments, often for ideological or political reasons.

America's *National Strategy for Homeland Security* defines terrorism as follows:

> Any premeditated, unlawful act dangerous to human life or public welfare that is intended to intimidate or coerce civilian populations or governments. (White House 2006)

The U.S. State Department defines terrorism thus:

> Premeditated, politically motivated violence perpetrated against noncombatant targets by subnational groups or clandestine agents. (U.S. Congress 2005)

The FBI definition of terrorism is as follows:

> The unlawful use of force or violence against persons or property to intimidate or coerce a Government, the civilian population, or any segment thereof, in furtherance of political or social objectives. (Federal Bureau of Investigation 1999)

Note that the FBI divides terrorism into two categories: domestic (homegrown), involving groups operating in and targeting the United States without foreign direction; and international, involving groups that operate across international borders and/or have foreign connections.

Well, at this point the obvious question is: Do you now know what terrorism it? That is, can you definitely define it? If you can't define it, you are not alone—not even the U.S. government can definitively define it. Maybe we need to look at other sources—views from the real experts on terrorism.

Osama bin Ladin's View on Terrorism:

> Wherever we look, we find the U.S. as the leader of terrorism and crime in the world. The U.S. does not consider it a terrorist act to throw atomic bombs at nations thousands of miles away [Japan during World War II], when those bombs would hit more than just military targets. Those bombs rather were thrown at entire nations, including women, children, and elderly people. (Bergen 2002, 21–22)

Another View

This view is from court testimony on terrorism from Ramzi Ahmed Yousef, who helped organize the first terrorist attack on the World Trade Center:

> You keep talking also about collective punishment and killing innocent people to force governments to change their policies; you call this terrorism when someone would kill innocent people or civilians in order to force the government to change its policies. Well, when you were the first one who invented this. . . .
>
> You were the first one who killed innocent people, and you are the first one who introduced this type of terrorism to the history of mankind when you dropped an atomic bomb

> which killed tens of thousands of women and children in Japan and when you killed over a hundred thousand people, most of them civilians, in Tokyo with fire bombings.
>
> You killed them by burning them to death. And you killed civilians in Vietnam with chemicals as with the so-called Orange agent. You killed civilians and innocent people, not soldiers, innocent people every single war you went. You went to wars more than any other country in this century, and then you have the nerve to talk about killing innocent people.
>
> And now you have invented new ways to kill innocent people. You have so-called economic embargo which kills nobody other than children and elderly people, and which other than Iraq you have been placing the economic embargo on Cuba and other countries for over 35 years. . . .
>
> The government in its summations and opening said that I was a terrorist. Yes, I am a terrorist and I am proud of it. And I support terrorism so long as it was against the United States Government and Israel, because you are more than terrorists; you are the one who invented terrorism and using it every day. You are butchers, liars and hypocrites. (Excerpt 1998)

And finally, here is an old cliché on a terrorist:

> One man's terrorist is another man's freedom fighter.

Again, from the preceding points of view, it can be seen that defining terrorism or the terrorist is not straightforward and never easy. Even the standard dictionary definition leaves us with the vagaries and ambiguities of other words typically associated with terrorism, such as in the definitions of *unlawful* and *public welfare* (Sauter and Carafano 2005).

Raphael Perl (2004) in a Congressional Research Service report points out that one definition widely used in U.S. government circles, and incorporated into law, defines *international terrorism* as terrorism involving the citizens or property of more than one country. *Terrorism* is broadly defined as politically motivated violence perpetrated against noncombatant targets by subnational groups or clandestine agents. For example, the kidnapping of U.S. birdwatchers or bombing of U.S.-owned oil pipelines by leftist guerrillas in Colombia would qualify as international terrorism. In 22 USC 2656f, a *terrorist group* is defined as a group that practices terrorism or has significant subgroups that practice terrorism. Perl (2004) points out that one of the shortfalls of this traditional definition is its focus on groups and its exclusion of individual ("lone wolf") terrorist activity, which has recently risen in frequency and visibility.

At this point, readers may wonder, "Why should we care?— that is, what difference does it make what the definition of terrorist or terrorism is?" Definitions are important because in order to prepare for the terrorism contingency, domestic or international, we must have some feel, as with any other problem, for what it is we are dealing with. We are fighting a war of ideas. We must attempt to understand both sides of the argument, even though the terrorist's side makes no sense to an American or other freedom-loving occupant of the globe.

Finally, while it is difficult to pinpoint an exact definition of terrorism, we certainly have little difficulty in identifying it when we see it, when we feel it, when we suffer from it. Consider, for example, in the earlier account of Willy's actions and the Admiral Sugar Processing Plant disaster. Put yourself in the place of those coworkers of his who were simply working to support themselves and their families. In particular, put yourself in the place of one of those female workers who one minute was a healthy, normal-looking young woman and then after the deflagration and countless hours of medical treatment, had her looks and health destroyed or compromised. When the event occurred, none of the victims could have known that an American terrorist had caused the act of terrorism on U.S. soil that killed eleven of his coworkers and badly maimed and disfigured forty others. No, they did not know that. However, there is one thing they knew for certain; they knew that crushing feeling of terror as they struggled to breathe, to survive. By any other name terrorism is best summed up as an absolute feeling of Terror—nothing judgmental about that—just Terror with a capital *T*.

VOCABULARY OF HATE

> In America, there are plenty of hate groups that claim peace and brotherhood, but when their actions are responsible for death and destruction, they are identified for what they really are. (Lindsey 2001)

After 9/11 several authors published and the media transmitted seemingly endless accounts of various hate groups operating throughout the world. Overnight Americans became aware of various theories, philosophies, and terminology very few had ever heard of or thought about. This trend is ongoing—never-ending.

Various pundits, so-called experts on the "new" genre of terrorism, have stated that for Americans to understand why foreign terrorists behead innocent people (or anyone else, for that matter) on television or blow up hospitals full of the sick or wounded or schoolhouses full of children, they must get inside the mind of a terrorist—enter the confines of the black box.

The average American might ask: "Get inside the mind of a terrorist? How the hell do you get inside the mind of a madman?"

This is where we make our first mistake, thinking that terrorists act in the manner they do because they are mad, nonrational, disturbed, or psychotic. In the case of Timothy McVeigh, we might be able to characterize him and his actions in this manner. Yet, again, McVeigh is the exception that proves the rule—terrorists' attacks, by real terrorists, are primarily planned beforehand by a group. It is important to remember that McVeigh acted primarily alone.

The terrorists that crashed airplanes into the Twin Towers, the Pentagon, and that farm field in Pennsylvania were all of the same mindset; they worked as a group. Likewise, the terrorists that attack Baghdad every day work as a group. Terrorists that did all the damage in Bali and Spain and elsewhere acted as a group. Thus, though we would like to classify all terrorists as we classify Timothy McVeigh, we can't do that. One madman working alone is something we can reasonably assume. However, thinking that hundreds or thousands of like-minded madmen all work in groups is a stretch—even though it is true. The cold-blooded manner in which terrorists go about their business suggests that they are not crazy, insane, or mad, but instead extremely harsh and calculating. If we dismiss them as madmen, we underestimate their intelligence. When we do that, we lose. No, we cannot underestimate the enemy—the terrorists. They are smart, cold-blooded, and calculating. In order to protect our critical chemical manufacturing infrastructure, we must be smarter and expect the unexpected—we must be proactive and not just reactive in implementing our countermeasures. Understanding is important. For example, understand that the Koran does not condone murder and suicide; instead, it provides injunctions against suicide and murder. When the Muslim terrorist commits murder or suicide in the name of Islam and the Koran, what he or she is really doing is changing the meaning of the words murder/suicide to mean martyrdom. No, getting into the mind of a terrorist is not the solution. Instead, in regard to terrorists, we must ensure they do not get into our minds. Simply, we must be smarter than the enemy.

The Language of Terrorism

Anyone who is going to work at improving the security of America's critical infrastructure must be well versed in the goals and techniques used by terrorists. Moreover, we cannot implement effective countermeasures unless we know our vulnerabilities. Along with this, we must not only understand what terrorists are capable of doing but also have some feel for their language or vocabulary, which will help us to understand where they are coming from and where they might be headed, so to speak.

As with any other technical presentation, understanding the information presented is difficult unless a common vocabulary is established. Voltaire said it best: "If you wish to converse with me, please define your terms." It is difficult enough to understand terrorists and terrorism; thus, we must be familiar with terms they use and that are used to describe them, their techniques, and their actions.

Definition of Terms

Abu Sayyaf—Meaning "bearer of the Sword"; the smaller of the two Islamist groups whose goal is to establish an Iranian-style Islamic state in Mindanao in the southern Philippines. In 1991, the group split from the Maro National Liberation Front with ties to numerous Islamic fundamentalist groups; they finance their operations through kidnapping for ransom, extortion, piracy, and other criminal acts. It is also thought that they receive funding from al Qaeda. It is estimated that there are between two hundred and five hundred Abu Sayyaf terrorists, mostly recruited from high schools and colleges.

acid bomb—A crude bomb made by combining muriatic acid with aluminum strips in a two-liter soda bottle.

aerosol—A fine mist or spray, which contains minute particles.

aflatoxin—A toxin created by bacteria that grow on stored foods, especially on rice, peanuts, and cottonseed.

Afghanistan—At the time of 9/11, governed by the Taliban and home of Osama bin Laden. Amid U.S. air strikes, which began on October 7, 2001, the U.S. sent in more than $300 million in humanitarian aid. In December 2001, Afghanistan reopened its embassy in the United States for the first time in more than twenty years.

agency—A division of government with a specific function, or a nongovernmental organization (e.g., private contractor, business, etc.) that offers a particular kind of assistance. In the incident command system, agencies are defined as jurisdictional (having statutory responsibility for incident mitigation) or assisting and/or cooperating (providing resources and/or assistance).

air marshal—A federal marshal whose purpose is to ride commercial flights in plainclothes and armed to prevent hijackings. Israel's use of air marshals on El Al is credited as the reason Israel has not had a single hijacking in thirty-one years. The United States started using air marshals after September 11. Despite President Bush's urging, there are not enough air marshals to go around, so many flights do not have them.

airborne—Carried by or through the air.

al-Gama'a al-Islamiyya (The Islamic Group, IG)—Islamic terrorist group that emerged spontaneously during the 1970s in Egyptian jails and later in Egyptian universities. After President Sadat released most of the Islamic prisoners from prisons in 1971, militants organized themselves in groups and cells, and al-Gama'a al-Islamiyya was one of them.

al Jazeera—Satellite television station based in Qatar and broadcast throughout the Middle East; often called the CNN of the Arab world.

al Qaeda—Meaning "the Base," an international terrorist group founded in approximately 1989 and dedicated to opposing non-Islamic governments with force and violence. One of the principal goals of al Qaeda was to drive the U.S. armed forces out of the Saudi Arabian peninsula and Somalia by violence. Currently wanted for several terrorist attacks, including those on the U.S. embassies in Kenya and Tanzania, as well as the first and second World Trade Center bombings and the attack on the Pentagon.

al Tahwid—A Palestinian group based in London, which professes a desire to destroy both Israel and the Jewish people throughout Europe. Eleven al Tahwid were arrested in Germany as they were allegedly about to begin attacking that country.
alpha radiation—The least penetrating type of nuclear radiation. Not considered dangerous unless particles enter the body.
American Airlines Flight 11—The Boeing 767 carrying eighty-one passengers, nine flight attendants, and two pilots, which was hijacked and crashed into the North Tower of the World Trade Center at 8:45 a.m. Eastern Time on September 11, 2001. Flight 11 was en route to Los Angeles from Boston.
American Airlines Flight 77—The Boeing 757 carrying fifty-eight passengers, four flight attendants, and two pilots, which was hijacked and crashed into the Pentagon at 9:40 a.m. Eastern Time on September 11, 2001. Flight 77 was en route to Los Angeles from Dulles International Airport in Virginia.
ammonium nitrate–fuel oil (ANFO)—A powerful explosive made by mixing fertilizer and fuel oil. The type of bomb used in the first World Trade Center attack, as well as the Oklahoma City bombing.
analyte—The name assigned to a substance or feature that describes it in terms of its molecular composition, taxonomic nomenclature, or other characteristic.
anthrax—An often fatal infectious disease contracted from animals. Anthrax spores have a long survival period, the incubation period is short, and disability is severe, making anthrax a bioweapon of choice for several nations.
antidote—A remedy to counteract the effects of poison.
antigen—A substance that stimulates an immune response by the body's immune system, which recognizes such a substance as foreign and produces antibodies to fight it.
antitoxin—An antibody that neutralizes a biological toxin.
Armed Islamic Group (GIA)—An Algerian Islamic extremist group, which aims to overthrow the secular regime in Algeria and replace it with an Islamic state. The GIA began its violent activities in early 1992, after Algiers voided the victory of the largest Islamic party, Islamic Salvation Front (FIS), in the December 1991 elections.
asymmetric threat—The use of crude or low-tech methods to attack a superior or more high-tech enemy.
axis of evil—Iran, Iraq, and North Korea, as mentioned by President George W. Bush during his State of the Union speech in 2002 as nations that were a threat to U.S. security due to harboring terrorism.
bioaccumulative—Substances that become concentrated in living organisms as they breathe contaminated air, drink or live in contaminated water, or eat contaminated food, rather than being eliminated through natural processes.

Biosafety Level 1—Suitable for work involving well-characterized biological agents not known to consistently cause disease in healthy adult humans, and of minimal potential hazard to lab personnel and the environment. Work is generally conducted on open benchtops using standard microbiological practices.

Biosafety Level 2—Suitable for work involving biological agents of moderate potential hazard to personnel and the environment. Lab personnel should have specific training in handling pathogenic agents and be directed by competent scientists. Access to the lab should be limited when work is being conducted, extreme precautions should be taken with contaminated sharp items, and certain procedures should be conducted in biological safety cabinets or other physical containment equipment if there is a risk of creating infectious aerosols or splashes.

Biosafety Level 3—Suitable for work done with indigenous or exotic biological agents that may cause serious or potentially lethal disease as a result of exposure by inhalation. Lab personnel must have specific training in handling pathogenic and potentially lethal agents and be supervised by competent scientists who are experienced in working with these agents. All procedures involving the manipulation of infectious material are conducted within biological safety cabinets or other physical containment devices, or by personnel wearing appropriate personal protective clothing and equipment. The lab must have special engineering and design features.

Biosafety Level 4—Suitable for work with the most infectious biological agents. Access to the two Biosafety Level 4 labs in the United States is highly restricted.

Bioterrorism Act—The Public Health Security and Bioterrorism Preparedness and Response Act of 2002.

BWC—Officially known as the Convention on the Prohibition of Development, Production, and Stockpiling of Bacteriological (Biological) and Toxin Weapons and on Their Destruction. The BWC works toward general and complete disarmament, including the prohibition and elimination of all types of weapons of mass destruction.

Baath Party—The official political party in Iraq until the United States debaathified Iraq in May 2003, after a war that lasted a little over a month. Saddam Hussein, the former ruler of the Baath Party, was targeted by American-led coalition forces and fled. Baath Party members have been officially banned from participating in any new government in Iraq.

Beltway sniper—For nearly a month in October 2002, the Washington D.C., Maryland, and Virginia area was the hunting grounds for forty-one-year-old John Allen Muhammad and seventeen-year-old Lee Boyd Malvo. Dubbed "the Beltway sniper" by the media, they shot people at seemingly random places such as schools, restaurants, and gas stations.

biochemical warfare—Collective term for the use of both chemical warfare and biological warfare weapons.

biochemterroism—Terrorism using biological or chemical agents as weapons.

biological ammunition—Ammunition designed specifically to release a biological agent used as the warhead for biological weapons. Biological ammunition may take many forms, such as a missile warhead or bomb.

biological attacks—The deliberate release of germs or other biological substances that cause illness.

bioterrorism—The use of biological agents in a terrorist operation. Biological toxins include anthrax, ricin, botulism, the plague, smallpox, and tularemia.

biowarfare—The use of biological agents to cause harm to targeted people either directly, by bringing the people into contact with the agents, or indirectly, by infecting other animals and plants, which would in turn cause harm to the people.

blister agents—Agents that cause pain and incapacitation instead of death and might be used to injure many people at once, thereby overloading medical facilities and causing fear in the population. Mustard gas is the best-known blister agent.

blood agents—Agents based on cyanide compounds. More likely to be used for assassination than for terrorism.

botulism—An illness caused by the botulinum toxin, which is exceedingly lethal and quite simple to produce. It takes just a small amount of the toxin to destroy the central nervous system. Botulism may be contracted by the ingestion of contaminated food or through breaks or cuts in the skin. Food supply contamination or aerosol dissemination of the botulinum toxin are the two ways most likely to be used by terrorists.

Bush Doctrine—The policy that holds responsible nations that harbor or support terrorist organizations and says that such countries are considered hostile to the United States. From President Bush's speech, "A country that harbors terrorists will either deliver the terrorists or share in their fate. . . . People have to choose sides. They are either with the terrorists, or they're with us."

Camp X-Ray—The Guantanamo Bay, Cuba, prison, which houses al Qaeda and Taliban prisoners.

carrier—A person or animal that is potentially a source of infection by carrying an infectious agent without visible symptoms of the disease.

cascading event—The occurrence of one event that causes another event.

causative agent—The pathogen, chemical, or other substance that is the cause of disease or death in an individual.

cell—The smallest unit within a guerrilla or terrorist group. A cell generally consists of two to five people dedicated to a terrorist cause. The formation of cells is born of the concept that an apparent "leaderless resistance" makes it hard for counterterrorists to penetrate.

chain of custody—The tracking and documentation of physical control of evidence.

chemical agent—A toxic substance (such as sarin or VX nerve gas) intended to be used for operations to debilitate, immobilize, or kill military or civilian personnel.

chemical ammunition—A munition, commonly a missile, bomb, rocket, or artillery shell, designed to deliver chemical agents.

chemical attack—The intentional release of a toxic liquid, gas, or solid in order to poison the environment or people.

chemical warfare—The use of toxic chemicals as weapons (not including herbicide used to defoliate battlegrounds or riot-control agents such as gas or mace).

chemical weapons—Weapons that produce effects on living targets via toxic chemical properties (e.g., sarin, VX nerve gas, and mustard gas).

chemterrorism—The use of chemical agents in a terrorist operation.

choking agent—Compound that injures primarily in the respiratory tract (i.e., nose, throat, and lungs). In extreme cases membranes swell up, lungs become filled with liquid, and death results from lack of oxygen.

Cipro—A Bayer antibiotic that combats inhalation anthrax.

"confirmed"—In the context of the threat evaluation process, a water contamination incident is characterized as "confirmed" if definitive evidence is found that the water has been contaminated.

counterterrorism—Measures used to prevent, preempt, or retaliate against terrorist attacks.

"credible"—In the context of the threat evaluation process, a water contamination threat is characterized as "credible" if information collected during the threat evaluation process corroborates information from the threat warning.

cutaneous—Related to or entering through the skin.

cutaneous anthrax—Antrax that is contracted via broken skin. The infection spreads through the bloodstream causing cyanosis, shock, sweating, and finally death.

cyanide agent— A colorless liquid that is inhaled in its gaseous form, or liquid cyanide and cyanide salts, which are absorbed by the skin. Used by Iraq in the Iran war against the Kurds in the 1980s, and also by the Nazis in the gas chambers of concentration camps. Symptoms are headache, palpitations, dizziness, and respiratory problems, followed by vomiting, convulsions, respiratory failure, unconsciousness, and eventually by death.

cyberterrorism—Attacks on computer networks or systems, generally by hackers working with or for terrorist groups. Some forms of cyberterrorism include denial-of-service attacks, inserting viruses, or stealing data.

dirty bomb—A makeshift nuclear device that is created from radioactive nuclear waste material. While not a nuclear blast, an explosion of a dirty bomb causes localized radioactive contamination as the nuclear waste material is carried into the atmosphere and dispersed by the wind.
Ebola—Ebola hemorrhagic fever (Ebola EF) is a severe, often-fatal disease in nonhuman primates such as monkeys, chimpanzees, and gorillas, and in humans. Ebola has appeared sporadically since 1976 when it was first recognized.
eBomb (for e-bomb)—Electromagnetic bomb, which produces a brief pulse of energy that affects electronic circuitry. At low levels, the pulse temporarily disables electronics systems, including computers, radios, and transportation systems. High levels completely destroy circuitry, causing mass disruption of infrastructure while sparing life and property.
ecoterrorism—Sabotage intended to hinder activities that are considered damaging to the environment.
Euroterrorism—Associated with left-wing terrorism of the 1960s, 1970s, and 1980s involving the Red Brigade, Red Army Faction, and November 17th Group, among other groups that targeted American interests in Europe and NATO. Other groups include Orange Volunteers, Red Hand Defenders, Continuity IRA, Loyalist Volunteer Force, Ulster Defense Association, and First of October Anti-Fascist Resistance Group.
fallout—The descent to the earth's surface of particles contaminated with radioactive material from a radioactive cloud. The term can also be applied to the contaminated particulate matter itself.
Fatah—Meaning "conquest by means of jihad," a political organization created in the 1960s and led by Yasser Arafat. With both a military and intelligence wing, it has carried out terrorist attacks on Israel since 1965. It joined the Palestine Liberation Organization (PLO) in 1968. Since 9/11, Fatah was blamed for attempting to smuggle fifty tons of weapons into Israel.
fatwa—A legal ruling regarding Islamic law.
Fedayeen Saddam—Iraq's paramilitary organization, which was said to be equivalent to the Nazis' SS. The militia was loyal to Saddam Hussein and was responsible for using brutality on civilians who were not loyal to the policies of Saddam. They did not dress in uniform.
filtrate—In ultrafiltration, the water that passes through the membrane and contains particles smaller than the molecular weight cutoff of the membrane.
frustration-aggression hypothesis—A hypothesis that every frustration leads to some form of aggression and every aggressive act results from some prior frustration. As defined by Gurr (1968), "The necessary precondition for violent civil conflict is relative deprivation, defined as actors' perception of discrepancy between their value expecta-

tions and their environment's apparent value capabilities. This deprivation may be individual or collective."

fundamentalism—Conservative religious authoritarianism. Fundamentalism is not specific to Islam; it exists in all faiths. Characteristics include literal interpretation of scriptures and a strict adherence to traditional doctrines and practices.

Geneva Protocol 1925—The first treaty to prohibit the use of biological weapons. Its full title is The 1925 Geneva Protocol for the Prohibition of the Use in War of Asphyxiating, Poisonous or Other Gases and of Bacteriological Methods of Warfare.

germ warfare—The use of biological agents to cause harm to targeted people either directly, by bringing the people into contact with the agents, or indirectly, by infecting other animals and plants, which would in turn cause harm to the people.

glanders—An infectious bacterial disease known to cause inflammation in horses, donkeys, mules, goats, dogs, and cats. Human infection has not been seen since 1945, but because so few organisms are required to cause disease, it is considered a potential agent for biological warfare.

grab sample—A single sample collected at a particular time and place that represents the composition of the water, air, or soil only at that time and location.

ground zero—From 1946 until 9/11, ground zero was the point directly above, below, or at which a nuclear explosion occurred, or the center or origin of rapid, intense, or violent activity or change. After 9/11, the term *Ground Zero* refers to the ground at the epicenter of the World Trade Center attacks.

guerrilla warfare— Literally, "little war." A term invented to describe the tactics Spain used to resist Napoleon, though the tactic itself has been around much longer. Guerilla warfare features cells and utilizes no front line. The oldest form of asymmetric warfare, guerilla warfare is based on sabotage and ambush with the objective of destabilizing the government through lengthy and low-intensity confrontation.

Hamas—A radical Islamic organization that operates primarily in the West Bank and Gaza Strip, whose goal is to establish an Islamic Palestinian state in place of Israel. On the one hand, Hamas operates overtly in its capacity as deliverer of social services, but its activists have also conducted many attacks, including suicide bombings, against Israeli civilians and military targets.

hazard—An inherent physical or chemical characteristic that has the potential for causing harm to people, the environment, or property.

hazard assessment—The process of evaluating available information about a site to identify potential hazards that might pose a risk to the site characterization team. The hazard assessment results in assigning one of four levels to risk: lower hazard, radiological hazard, high chemical hazard, or high biological hazard.

hemorrhagic fever, or viral hemorrhagic fever—A term used to describe a severe multisystem syndrome, wherein the overall vascular system is damaged and the body becomes unable to regulate itself. These symptoms are often accompanied by hemorrhage; however, the bleeding itself is not usually life-threatening. Some types of hemorrhagic fever viruses can cause relatively mild illnesses.

Hizbollah (Hezbollah)—Meaning "The Party of God." One of many terrorist organizations that seek the destruction of Israel and of the United States. It has taken credit for numerous bombings against civilians, and has declared that civilian targets are warranted. Hezbollah claims it sees no legitimacy for the existence of Israel, and that their conflict is one of legitimacy that is based on religious ideals.

Homeland Security Office—An agency organized after 9/11, with former Pennsylvania Governor Tom Ridge heading it up. The Office of Homeland Security is at the top of approximately forty federal agencies charged with protecting the United States against terrorism.

homicide bombings—A term coined by the White House to replace the old "suicide bombings."

incident—A confirmed occurrence that requires response actions to prevent or minimize loss of life or damage to property and/or natural resources. For example, a drinking water contamination incident occurs when the presence of a harmful contaminant has been confirmed.

inhalation anthrax—A form of anthrax that is contracted by inhaling anthrax spores. This results in pneumonia, sometimes meningitis, and finally death.

intifada (intifadah)—From Arabic, meaning "shaking off." There have been two intifadas, which were similar in that both were originally characterized by civil disobedience by Palestinians that escalated into the use of terror. In 1987, following the killing of several Arabs in the Gaza Strip, the first intifada began and went on until 1993. The second intifada began in September 2000, following Ariel Sharon's visit to the Temple Mount.

Islam—Meaning "submit." The faith practiced by followers of Muhammad, Islam claims more than a billion believers worldwide.

jihad—Meaning "struggle." The definition is a subject of vast debate; there are two definitions generally accepted. The first is struggle against oppression, whether political or religious. The second is struggle within oneself, or a spiritual struggle.

kneecapping—A malicious wounding by firearm to damage the knee joint; a common punishment used by Northern Ireland's IRA for those collaborating with the British.

Koran (or Qu'ran)—The holy book of Islam, considered by Muslims to contain the revelations of God to Muhammad.

LD50—The dose of a toxic substance that kills 50 percent of those infected.

Laboratory Response Network (LRN)—A network of labs developed by the CDC, APHL, and FBI for the express purpose of dealing with bioterrorism threats, including pathogens and some biotoxins.

lassa fever—An acute, often fatal, viral disease characterized by high fever, ulcers of the mucous membranes, headaches, and disturbances of the gastrointestinal system.

link—The means (road, rail, barge, or pipeline) by which a chemical is transported from one node to another.

mind-set—According to *American Heritage Dictionary*, "1. A fixed mental attitude or disposition that predetermines a person's response to and interpretation of situations; 2. An inclination or a habit." *Merriam Webster's Collegiate Dictionary* (11th ed.) defines it as "1. A mental attitude or inclination; 2. a fixed state of mind." The term dates from 1909 but apparently is not included in dictionaries of psychology.

Molotov cocktail—A crude incendiary bomb made of a bottle filled with flammable liquid and fitted with a rag wick.

monkeypox—A virus in the same family as smallpox, used in the Russian bioweapons program. In June 2003, a spate of human monkeypox cases was reported in the U.S. Midwest. This was the first time that monkeypox was seen in North America, and it was the first time that monkeypox was transferred from animal to human. There was some speculation that it was a bioattack.

mullah—A Muslim, usually holding an official post, who is trained in traditional religious doctrine and law.

Muslim (also Moslem)—Followers of the teachings of Muhammad, or Islam.

mustard gas—A blistering agent that causes severe damage to the eyes, internal organs, and respiratory system. Produced for the first time in 1822, mustard gas was not used until World War I. Victims suffered the effects of mustard gas thirty to forty years after exposure.

narcoterrorism—The view of many counterterrorist experts that there exists an alliance between drug traffickers and political terrorists.

National Pharmaceutical Stockpile—A stock of vaccines and antidotes stored at the Centers for Disease Control and Prevention in Atlanta, to be used against biological warfare.

nerve agent—First used by the Nazis, the first nerve agents were insecticides developed into chemical weapons. Some of the better known nerve agents include VX, sarin, soman, and tabun. These agents are used because only a small quantity is necessary to inflict substantial damage. Nerve agents can be inhaled or can be absorbed through intact skin.

node—A facility at which a chemical is produced, stored, or consumed.

nuclear blast—An explosion of any nuclear material, which is accompanied by a pressure wave, intense light and heat, and widespread radioactive fallout, which can contaminate the air, water, and ground surface for miles around.

opportunity contaminant—A contaminant that may be readily available in a particular area, even though it may not be highly toxic or infectious or easily dispersed and stable in treated drinking water.

Osama (or Usama) bin Laden—A native of Saudi Arabia, the seventeenth of twenty-four sons of Saudi Arabian builder Mohammed bin Oud bin Laden, a Yemeni immigrant. Early in his career, bin Laden helped the mujahideen fight the Soviet Union by recruiting Arabs and building facilities. He hates the United States, and apparently this is because he views the United States as having desecrated holy ground in Saudi Arabia with their presence during the first Gulf War. Expelled from Saudi Arabia in 1991 and from Sudan in 1996, he operated terrorist training camps in Afghanistan. His global network, al Qaeda, is credited with the attacks on the United States on September 11, 2001, the attack on the USS *Cole* in 2000, and numerous other terrorist attacks.
pathogen—Any agent that can cause disease.
pathway—The sequence of nodes and links by which a chemical is produced, transported, and transformed, from its initial source to its ultimate consumer.
plague—The pneumonic plague, which is more likely to be used in connection with terrorism. It is naturally carried by rodents and fleas but can be aerosolized and sprayed from crop dusters. A 1970 World Health Organization assessment asserted that, in a worst-case scenario, a dissemination of fifty kilograms in an aerosol over a city of five million could result in 150,000 cases of pneumonic plague, 80,000–100,000 of whom would require hospitalization, and 36,000 of whom would be expected to die.
political terrorism—Terrorist acts directed at governments and their agents and motivated by political goals (e.g., national liberation).
"possible"—In the context of the threat evaluation process, a water contamination threat is characterized as "possible" if the circumstances of the threat warning appear to have provided an opportunity for contamination.
potassium iodide—An FDA-approved nonprescription drug for use as a blocking agent to prevent the thyroid gland from absorbing radioactive iodine.
presumptive results—Results of chemical and/or biological field testing that need to be confirmed by further lab analysis. Typically used in reference to the analysis of pathogens.
psychopath—A mentally ill or unstable person, especially one having a psychopathic personality, according to *Webster's*.
psychopathy—A mental disorder, especially an extreme mental disorder marked usually by egocentric and antisocial activity, according to *Webster's*.
psychopathology—The study of psychological and behavioral dysfunction occurring in mental disorder or in social disorganization, according to *Webster's*.
psychotic—Of, relating to, or affected with psychosis, which is a fundamental mental derangement (such as schizophrenia) characterized by defective or lost contact with reality, according to *Webster's*.
rapid field testing—Analysis of water during site characterization in an attempt to tentatively identify contaminants or unusual water quality.

red teaming—As used in this text, a group exercise to imagine all possible terrorist attack scenarios against the chemical infrastructure and their consequences.
retentate—In ultrafiltration, the solution that contains the particles that do not pass through the membrane filter. The retentate is also called the concentrate.
ricin—A stable toxin easily made from the mash that remains after castor beans are processed. At one time, it was used as an oral laxative—castor oil. Ricin causes diarrhea, nausea, vomiting, abdominal cramps, internal bleeding, liver and kidney failure, and circulatory failure. There is no antidote.
salmonella enteritis—An infection caused by a gram-negative bacillus, a germ of the Salmonella genus. Infection with this bacteria may involve only the intestinal tract or may be spread from the intestines to the bloodstream and then to other sites in the body. Symptoms include diarrhea, nausea, fever, and abdominal cramps. Dehydration resulting from the diarrhea can cause death, and the disease could cause meningitis or septicemia. The incubation period is between eight and forty-eight hours, while the acute period of the illness can hang on for one to two weeks.
sarin—A colorless, odorless gas. With a lethal dose of .5 mg (a pinprick-sized droplet), it is twenty-six times more deadly than cyanide gas. Because the vapor is heavier than air, it hovers close to the ground. Sarin degrades quickly in humid weather, but sarin's life expectancy increases as the temperature gets higher, regardless of how humid it is.
sentinel laboratory—An LRN lab that reports unusual results that might indicate a possible outbreak, and refers specimens that may contain select biological agents to reference labs within the LRN.
site characterization—The process of collecting information from an investigation site in order to support the evaluation of a drinking water contamination threat. Site characterization activities include site investigation, field safety screening, rapid field testing of the water, and sample collection.
sleeper cell—A small cell that keeps itself undetected until such time as it can "awaken" and cause havoc.
smallpox—The first biological weapon, used during the eighteenth century, smallpox killed three hundred million people in the nineteenth century. There is no specific treatment for smallpox, and the only prevention is vaccination. This currently poses a problem, since the vaccine was discontinued in 1970 and the World Health Organization (WHO) declared smallpox eradicated. Incubation is seven to seventeen days, during which the carrier is not contagious. Thirty percent of people exposed are infected, and it has a 30 percent mortality rate.
sociopath—Basically synonymous with *psychopath* (see above). Symptoms in the adult sociopath include an inability to tolerate delay or frustration, a lack of guilt feelings, a relative lack of anxiety, a lack of compassion for others, a hypersensitivity to personal

ills, and a lack of responsibility. Many authors prefer the term sociopath because this type of person had defective socialization and a deficient childhood.

sociopathic—Of, relating to, or characterized by asocial or antisocial behavior or a psychopathic personality, according to *Webster's*.

spore—An asexual, usually single-celled reproductive body of plants such as fungi, mosses or ferns; a microorganism, as a bacterium, in a resting or dormant state.

terrorist group—A group that practices terrorism or has significant elements that are involved in terrorism.

threat—An indication that a harmful incident, such as contamination of the drinking water supply, may have occurred. The threat may be direct, such as a verbal or written threat, or circumstantial, such as a security breach or unusual water quality.

toxin—A poisonous substance produced by living organisms that is capable of causing disease when introduced into the body tissues.

transponder—A device on an airliner that sends out a signal allowing air traffic controllers to track the plane. Transponders were disabled in some of the planes hijacked on 9/11.

Transportation Security Administration (TSA)—A new agency created by the Patriot Act of 2001 for the purpose of overseeing technology and security in American airports.

tularemia—An infectious disease caused by the hardy bacterium *Francisella tularensis*, found in animals, especially rabbits, hares, and rodents. Symptoms depend upon how the person was exposed to tularemia but can include difficulty breathing, chest pain, bloody sputum, swollen and painful lymph glands, ulcers on the mouth or skin, swollen and painful eyes, and sore throat. Symptoms usually appear from three to five days after exposure but sometimes up to two weeks after exposure. Tularemia is not spread from person to person, so people who have it need not be isolated.

ultrafiltration—A filtration process for water that uses membranes to preferentially separate very small particles that are larger than the membrane's molecular weight cutoff, typically greater than 10,000 Daltons. (A Dalton is a unit of mass, defined as one-twelfth the mass of a carbon-12 nucleus. It is also called the *atomic mass unit*, abbreviated as either *amu* or *u*.)

vector—An organism that carries germs from one host to another.

vesicle—A blister filled with fluid.

weapons of mass destruction (WMD)—According to the National Defense Authorization Act, any weapon or device that is intended, or has the capability, to cause death or serious bodily injury to a significant number of people through the release, dissemination, or impact of

- toxic or poisonous chemicals or their precursors
- a disease organism
- radiation or radioactivity

xenophobia—Irrational fear of strangers or those who are different from oneself.
zyklon b—A form of hydrogen cyanide. Symptoms of inhalation include increased respiratory rate, restlessness, headache, and giddiness followed later by convulsions, vomiting, respiratory failure, unconsciousness, and death. Used in the Nazi gas chambers in World War II.

REFERENCES AND RECOMMENDED READING

Bergen, P. L. 2002. *Holy War, Inc.: Inside the secret world of Osama bin Ladin.* New York: Touchstone Press.

Excerpt from court testimony. 1998. *New York Times,* January 9, B4.

Federal Bureau of Investigation (FBI). 1999. Terrorism in the United States 1999. http://www.fbi.gov/publications/terror/terror99.pdf.

Gurr, T. R. 1968. Psychological factors in civil violence. *World Politics* 20 (32), 245–78.

Haimes, Y. Y. 2004. *Risk modeling, assessment, and management.* 2nd ed. New York: Wiley.

Henry, K. 2002. New face of security. *Government Security,* April, 30–37.

Lindsey, H. 2001. Vocabulary of hate. www.wordnetdaily.com (accessed April 18, 2008).

Meyer, E. 2004. *Chemistry of hazardous materials.* 4th ed. Upper Saddle River, NJ: Prentice Hall.

Occupational Safety and Health Administration (OSHA). 2007. Combustible Dust National Emphasis Program. CPL 03-00-006. www.osha.gov/pls/oshaweb/owadisp.show_document?p_table=DIRECTIVES&p_id=3729 (accessed April 14, 2008).

Perl, R. 2004. *Terrorism and national security: Issues and trends.* CRS Issue Brief IB10119. Washington, DC: Congressional Research Service.

Sauter, M. A., and J. J. Carafano. 2005. *Homeland security: A complete guide to understanding, preventing, and surviving terrorism.* New York: McGraw-Hill.

Shea, D. A. 2008. *Chemical facility security: Regulation and issues for Congress.* Washington, DC: Congressional Research Service.

Spellman, F. R. 1997. *A guide to compliance for process safety management/risk management planning (PSM/RMP).* Lancaster, PA: Technomic Publishing Company.

U.S. Congress. 2005. Annual country reports on terrorism. 22 USC, Chapter 38, Section 2656f.

White House. 2006. National strategy for homeland security. www.whitehouse/homeland (accessed May 13, 2006).

2

Critical Infrastructure

While it is not so easy to definitively define terrorism and/or the terrorist, we have less difficulty identifying the likely targets of terrorists. In America, we call these likely targets our critical infrastructure.

WHAT IS CRITICAL INFRASTRUCTURE?

For the United States of America, 9/11 was a slap in the face, a punch in the gut (actually, the ultimate sucker punch), and a most serious wake-up call. The 9/11 wake-up call generated several reactions on our part—obviously, protecting ourselves from further attack became (and hopefully still is) priority number one. In light of this important need (i.e., the survival of our way of life), the Department of Homeland Security was created. According to Barack Obama (2007), the Department of Homeland Security does "the work that ensures no other family members have to lose a loved one to a terrorist who turns a plane into a missile, a terrorist who straps a bomb around her waist and climbs aboard a bus, a terrorist who figures out how to set off a dirty bomb in one of our cities."

Among other things, the new emphasis on homeland security pointed to the need to protect and enhance the security of the nation's critical infrastructure. Critical infrastructure can be defined or listed in many ways. Generally, governments use the term to describe material assets that are essential for the functioning of an economy and a society. For the purpose of this text, critical infrastructure is defined as those assets of physical and computer-based systems that are essential to the minimum operations of our economy and government. Critical infrastructures (in the authors' opinion), are the following:

- agriculture
- banking and finance

- chemical and hazardous materials
- defense industrial base
- emergency services
- energy
- national monuments and icons
- organizations
- postal and shipping
- public health
- strategies and assessments
- telecommunications
- transportation
- water

Although we did not list cyberspace and all ancillaries involved in or with digital operations (e-technology), in this current era we can state without equivocation, doubt, ambiguity, or vagary that the digital connection is the glue that holds all critical infrastructure together. This is the case, of course, because all separate infrastructures are interconnected in one way or another. This may surprise you to some degree, but think about it—we are not speaking about rocket science here; we are speaking about the present reality of e-technology. It would be hard to imagine that any of the above-listed infrastructure sectors could operate today without e-technology.

For example, consider e-agriculture. The Food and Agriculture Organization of the United Nations (2005) defines e-agriculture as "an emerging field in the intersection of agricultural informatics, agricultural development and entrepreneurship, referring to agriculture services, technology dissemination, and information delivered or enhanced through the Internet and related technologies. More specifically, it involves the conceptualization, design, development, evaluation and application of new (innovative) ways to use existing or emerging information and communication technologies."

DID YOU KNOW?

More advanced applications of e-agriculture in farming exist in the use of sophisticated information and communication technologies such as satellite systems, Global Positioning Systems (GPS), and advanced computers and electronic systems to improve the quantity and quality of production (Food and Agriculture Organization of the United Nations 2005).

Even though we did it in the past, today how would we go about withdrawing money from the bank without e-banking technology? Today we can conduct our banking at any time we wish, from any location in the world. Modern life without debit cards and ATMs would be a rude awakening for many of us.

We could go down the list of critical infrastructures and easily point out where e-technology and the specific industry interface. However, since this is a discussion focusing on the chemical industry, we will keep the focus on e-technology as it relates to or interfaces with the chemical industry. If you are not familiar with the chemical industry—an oil refinery, for example—it might surprise you to know that the entire modern refining process is operated by e-technology from computer operation stations (staffed by people, of course) with various digital proximity switches and other devices strategically positioned throughout the process to operate valves, monitor critical parameters, and provide automatic emergency shutdown procedures.

CHEMICAL INDUSTRY INFRASTRUCTURE

In the first two volumes of our critical infrastructure series, *Water Infrastructure Protection and Homeland Security* and *Food Infrastructure Protection and Homeland Security*, the message and focus was (as the titles suggest) water/wastewater and agriculture. In this third volume of the series, *Chemical Infrastructure Protection and Homeland Security*, even though the target is different—pointing out and discussing the threat to our chemical industry—we use the same proven format. In addition, this text describes the study, design, and implementation of precautionary measures aimed at reducing the risk to our chemical industry from both homegrown and/or foreign terrorism.

Here we are concerned with chemical industry (including hazardous materials) infrastructure, which we define as follows:

> Chemical and hazardous materials industry infrastructure includes substantial facility and equipment investment; it is highly capital intensive. Most chemical industry facilities contain very specialized process equipment that would be difficult to replace quickly. A good example is an oil refinery plant, where if the cracking facilities were destroyed they could not be replaced anytime soon. It is interesting to note that some chemical industry facilities (e.g., oil refineries) require large amounts of land (have a large footprint) but are typically staffed with few employees relative to on-site land requirements.

Again, it is important to point out that the chemical industry provides products and materials that are essential to the U.S. economy and to the so-called "good life," the standard of living we presently enjoy. In addition to the economic consequences of a successful homegrown or foreign terrorist attack against chemical industry facilities, there is also the potential of a threat to public health and safety and the environment.

We hope that this book and the others in the critical infrastructure series will aid in the prevention and mitigation of deliberate attacks.

HSPD-7 (PROTECTING CRITICAL INFRASTRUCTURE)

In terms of protecting critical infrastructure, agriculture was added to the list in December 2003 by Homeland Security Presidential Directive 7 (HSPD-7), "Critical Infrastructure Identification, Prioritization, and Protection." This directive instructs agencies to develop plans to prepare for and counter the terrorist threat. HSPD-7 mentions the following industries: agriculture and food; banking and finance, transportation (air, sea, and land, including mass transmit, rail, and pipelines); energy (electricity, oil, and gas); telecommunications; public health; emergency services; drinking water; and water treatment (wastewater treatment is implied).

REFERENCES AND RECOMMENDED READING

Breeze, R. 2004. Agroterrorism: Betting far more than the farm. *Biosecurity and Bioterrorism: Biodefense Strategy, Practice and Science* 2 (4): 1–14.

Carus, S. 2002. *Bioterrorism and biocrimes: The illicit use of biological agents since 1900.* Washington, DC: Center of Counterproliferation Research, National Defense University.

Centers for Disease Control and Prevention (CDC). 2003. Nicotine poisoning after ingestion of contaminated ground beef—Michigan, 2003. *Morbidity and Mortality Weekly Report* 52 (18): 413–16.

———. 2006. Facts about ricin. www.bt.cdc.gov/agent/ricin/facts.asp (accessed June 27, 2007).

Chalk, P. 2004. *Hitting America's soft underbelly: The threat of deliberate biological attacks against the U.S. agriculture and food industry.* Santa Monica, CA: RAND Corp.

Collins, S. M. 2003. *Agroterrorism: The threat to America's breadbasket.* Washington, DC: U.S. Senate Committee on Governmental Affairs.

Congressional Budget Office (CBO). 2004. *Homeland security and the private sector.* Washington, DC: Congressional Budget Office.

Congressional Federal Register. 2003. Notice of proposed rulemaking. *Federal Register* 68, no. 90.

Congressional Research Service (CRS). 2003. Mad cow disease and U.S. beef trade. CRS Report for Congress. www.ers.usda.gov/features/fse/index.htm (accessed July 4, 2007).

Food and Agriculture Organization of the United Nations (FAO). 2005. *Bridging the rural digital divide.* New York: United Nations. www.fao.org/rdd (accessed April 19, 2008).

Food and Drug Administration (FDA). 2003. Risk assessment for food terrorism and other food safety concerns. www.fsan.fda.gov/~dms/rabtact.html (accessed June 27, 2007).

Government Accountability Office (GAO). 2003. Bioterrorism: A threat to agriculture and the food supply. www.gao.gov.htex/do4259.html (accessed June 10, 2007).

Henry, K. (2002). New face of security. *Government Security* (April): 30–37.

Horn, F. P. 1999. Statement made before the United States Senate Emerging Threats and Capabilities Subcommittee of the Armed Services Committee. www.Senate.gov/~armed_servies/statement/1999/991027fh.pdf (accessed June 27, 2007).

Knowles, T., J. Lane, G. Bayens, N. Speer, J. Jaax, D. Carter, and A. Bannister. 2005. *Defining law enforcement's role in protecting American agriculture from agroterrorism.* Washington, DC: U.S. Department of Justice. www.ncjrs.gov/pdffiles1/nij/grants (accessed July 7, 2007).

Lane, J. 2002. Sworn testimony, Congressional Field Hearing, House Committee on Government Reform, Abilene, Kansas.

McCoy, A. W. 1972. *The politics of heroin in Southeast Asia.* New York: Harper & Row.

Monke, J. 2004. Agroterrorism: Threats and preparedness. Washington, DC: Congressional Research Service.

Obama, B. 2007. Homeland security. http://www.whitehouse.gov/agenda/homeland_security/.

Parker, H. S. 2002. Agricultural bioterrorism: A federal strategy to meet the threat. McNair Paper 65, Nation Defense University. www.ndu.edu/inss/McNair/mcnair65/McN_65.

Ryan, C. A., M. K. Nickels, N. T. Hargrett-Bean, et al. 1987. Massive outbreak of antimicrobial-resistant salmonellosis traced to pasteurized milk. *Journal of the American Medical Association* 258 (22): 3269–74.

Spellman, F. R., and N. E. Whiting. 2007. *Environmental management of concentrated animal feeding operations (CAFO).* Boca Raton, FL: CRC Press.

U.S. Department of Agriculture (USDA). 2004. Economic Research Service. Agricultural Outlook tables. www.ers.usda.gov/publications/Agoutlook/AOTables.

———. 2007. Fact sheet: Interim melamine and analogues safety/risk assessment. Washington, DC: United States Department of Agriculture.

World Health Organization (WHO). 2002. Food safety issues: Terrorist threats to food. www.who.int/fsf (accessed June 26, 2007).

3

The Chemical Industry

Evidence that U.S. chemical plants may be used by terrorists to gain access to chemicals exists. For example, one of the 1993 World Trade Center bombers became a naturalized U.S. citizen, graduated from Rutgers University, and worked as a chemical engineer at Allied Signal, from which he used company stationery to order chemical ingredients to make the bomb. According to a U.S. prosecutor in the case against the bombers, though "some suppliers balked when the order came from outside official channels, when the delivery address was a storage park, or when [a co-conspirator] tried to pay for the chemicals in cash" (Parachini 2000), others did not. Moreover, testimony at the trial of the bombers indicated that they had successfully stolen cyanide from a chemical facility and were training to introduce it into the ventilation systems of office buildings (Parachini 2000). Bond (2002) points out that more recently, chemical trade publications reportedly were found in al Qaeda hideaways.

The U.S. chemical industry is vital to the U.S. economy. It is a high-tech, research and development (R&D)-oriented industry that is awarded about one out of every eight U.S. patents. The chemical industry is defined by Standard Industry Code (SIC) 28, Chemicals and Allied Products. Chemicals is a broad, complex, diverse, and wide-ranging industry that produces over seventy thousand different products. It is an essential component of manufacturing such as in construction, motor vehicles, paper, electronics, transportation, agriculture, and pharmaceuticals. Although some chemical manufacturers produce and sell consumer products such as bleach, cosmetics, and soap, most chemical products are used as intermediate products for other goods. The facilities in which chemicals are produced range from refineries covering square miles of land with many high-volume chemicals on-site, to much smaller operations. The chemical industry produces 1.9 percent of U.S. gross domestic product (GDP). It is the nation's number one exporter (U.S. Department of Commerce 1996; Bureau of Labor Statistics 2008). The U.S. Department of Commerce (1996) lists chemicals as a "keystone" industry—one critical to the global competitiveness of other U.S. industries.

According to the Bureau of Labor Statistics (2008), about 54 percent of those employed in the chemical industry work in production and in installation, maintenance, and repair occupations. Another 12 percent work in professional and related occupations. Approximately 9 percent work in management, business, and financial occupations and in office and administrative support occupations, and another 9 percent work in transporting and material-moving occupations.

DID YOU KNOW?

The U.S. chemical industry employs more than one million people and produces more than seventy thousand products.

MAJOR COMPONENTS OF THE U.S. CHEMICAL INDUSTRY (USDC 1996)

Because of the many different products and processes of the chemical industry it is difficult to make a meaningful description of the chemical industry. However, essentially at the base of the chemical industry are companies that combine organic and inorganic materials from the earth with heat, air, and water to make chemicals that, in turn, are essential to products used in everyday life in modern economies. Sidebar 3.1 outlines the major components of the industry.

Sidebar 3.1. Major Components of the U.S. Chemical Industry

The U.S. chemicals and allied products industry consists of some 9,125 corporations whose primary business is the development, manufacturing, and marketing of industrial chemicals, pharmaceuticals, and other chemical products.

The industrial chemicals segment (SICs 281, 282, and 286) of the industry consists of some 1,725 corporations whose primary business is the manufacturing and marketing of alkalis and chlorine, inorganic pigments, industrial gases, and other industrial inorganic chemicals; plastic resins, synthetic rubber, and man-made fibers; and petrochemicals and other industrial organic chemicals.

The pharmaceuticals segment (SIC 283) consists of some 1,225 corporations whose primary business is the development, manufacturing, and marketing of medicinal chemicals and botanicals; in vitro and other diagnostic substances to diagnose or monitor the state of human or veterinary health; bacterial and virus vaccines, toxoids, serums, plasmas, and other biological products for human and veterinary health; and vitamins and other pharmaceutical preparations for both human and veterinary use.

Other chemical products (SICs 284, 285, 287, and 289) consist of some 6,175 corporations whose primary business is the manufacturing and marketing of soaps and detergents; surfactants; specialty cleaning, polishing, and sanitary preparations; perfumes, cosmetics, and other toilet preparations; paints, varnishes, enamels, and other allied products; fertilizers, pesticides, and other agricultural chemicals; and adhesives and sealants, explosives, printing ink, and other specialty chemicals and chemical preparations.

Source: U.S. Department of Commerce (1996).

CHEMICAL MANUFACTURING, STORAGE, AND USE IN THE UNITED STATES

We use sidebar 3.2, an outline of the major categories of the chemical industry, to identify the nation's vulnerabilities to terrorist attack on our chemical infrastructure. Virtually all chemical use, storage, and manufacturing in the United States fits into one of these categories.

Sidebar 3.2. Major Chemical Categories

Petrochemicals and fossil fuels entail chemicals produced from hydrocarbon feedstocks, such as crude oil products and natural gas. They include such chemicals as hydrocarbons and industrial chemicals (e.g., alcohols, acrylates, acetates), aromatics (e.g., benzene, toluene, xylenes), and olefins (e.g., ethylene, propylene, butadiene, methanol).

Inorganic chemicals and fertilizers include acids (e.g., sulfuric, nitric) and alkalies (e.g., caustic soda, soda ash), chlorine, ammonia, and ammonia-derived fertilizers. They also include fluorine derivatives (e.g., hydrogen fluoride), phosphates, potash, pigments (e.g., titanium dioxide), and certain metals such as mercury.

Industrial gases encompass two general classes: (1) gases used primarily in large quantities as auxiliaries in other manufacturing processes (e.g., refining, petrochemical, or steel manufacture), including nitrogen, oxygen, hydrogen, and carbon monoxide, and (2) specialty gases that are produced in smaller quantities to serve the electronics, food, and other industries.

Specialty chemicals comprise a large number of chemicals that are used as aids to the manufacture of other major products (e.g., in paper milling, plastic production, water treatment, mining), are used as end products (e.g., pesticides in farming), or are components of consumer products (personal care products, paints and coatings, adhesives and sealants, photographic chemicals).

(*continued*)

Pharmaceuticals include prescription and over-the-counter drugs, diagnostic substances, vaccines, vitamins, and preparations for both human and veterinary uses.

Consumer products entail formulated products, such as soaps, detergents, bleaches, paints, solvents, glues, toothpaste, shampoos, cosmetics, skin care products, perfumes, and colognes intended for direct consumer use.

Source: Adapted from Board on Chemical Sciences and Technology (2006).

THE CHEMICAL INDUSTRY: A REGULATED ENVIRONMENT (USDC 1996)

Even before 9/11, the chemical industry was (and is) one of the nation's most regulated industries. It is subject to numerous environmental regulations as well as to the voluntary obligations imposed by the chemical industry's environmental, health, and safety improvement initiatives. Including federal/state OSHA statutes, fifteen major federal statutes, as well as numerous state laws, impose significant compliance and reporting requirements on the industry (see sidebar 3.3).

Sidebar 3.3. Major Health, Safety, and Environmental Legislation

1. **Toxic Substances Control Act (TSCA) of 1976** gives the Environmental Protection Agency (EPA) comprehensive authority to regulate any chemical substance whose manufacture, processing, distribution in commerce, use, or disposal may present an unreasonable risk of injury to health or the environment.
2. **Clean Air Act (CAA)** was first passed in 1955 as the Air Pollution Control, Research and Technical Assistance Act and amended in 1963 to become the CAA. A more significant statute was passed in 1970 and amended in 1977 and 1990. It provides EPA authority to regulate air pollutants from a wide variety of sources including automobiles, electric power plants, chemical plants, and other industrial sources.
3. **Clean Water Act (CWA)** was first enacted in 1948 as the Federal Water Pollution Control Act. Subsequent extensive amendments defined the statute to be known as the CWA in 1972; it was further amended in 1977 and 1987. The CWA provides EPA authority to regulate effluents from sewage treatment works, chemical plants, and other industry sources into U.S. waterways. EPA has recently undertaken control efforts in on-point source pollution as well.
4. **Comprehensive Environmental Response Compensation and Liability Act of 1980 (CERCLA)** and **the Superfund Amendments and Reauthorization Act of 1986 (SARA)** provide the basic legal framework for the federal "Superfund" program to clean up abandoned hazardous waste sites.
5. **Federal Insecticide, Fungicide, and Rodenticide Act (FIFRA)** provides EPA authority to register and assess the risks of agricultural pesticides. It was first enacted in 1947 and last amended in 1988.

6. **Federal Food, Drug and Cosmetics Act (FDCA)** provides the Food and Drug Administration (FDA) authority to regulate the manufacturing of drugs and pharmaceuticals and the use of packaging and additives in food and cosmetics.
7. **Emergency Planning and Community Right-to-Know Act of 1986**, also known as SARA Title III, mandates state and community development of emergency preparedness plans and also establishes an annual manufacturing-sector emissions reporting program.
8. **Resource Conservation and Recovery Act (RCRA) of 1976** provides the EPA with authority to establish standards and regulations for handling and disposing of solid and hazardous wastes.
9. **Occupational Safety and Health Act (OSH Act) of 1970** provides the Department of Labor authority to set comprehensive workplace safety and health standards, including permissible exposures to chemicals in the workplace, and authority to conduct inspections and issue citations for violations of safety and health regulations.
10. **Safe Drinking Water Act**, enacted in 1974 and amended in 1977 and again in 1986, establishes standards for public drinking water supplies.
11. **Hazardous Materials Transportation Act (HMTA)** provides the Department of Transportation the authority to regulate the packaging and movement of hazardous materials.
12. **Chemical Diversion and Trafficking Act (CDTA) of 1988** is designed to prevent the diversion of chemicals to illegal drug producers.
13. **Pollution Prevention Act of 1990** makes it the national policy of the United States to reduce or eliminate the generation of waste at the source whenever feasible and directs the EPA to undertake a multimedia program of information collection, technology transfer, and financial assistance to the states to implement this policy and to promote the use of source reduction techniques.
14. **Flammable Fabrics Act**, enacted in 1970 and last amended in 1983, gives the Consumer Product Safety Commission the authority to set flammability standards for fabrics that protect against an unreasonable risk of the occurrence of a fire.
15. **Poison Packaging Prevention Act of 1953**, last amended in 1990, provides the Consumer Product Safety Commission authority to set standards for the special packaging of any household product to protect children from a hazard.
16. **Consumer Product Safety Act**, enacted in 1972, created the Consumer Product Safety Commission and gives the Commission authority to issue mandatory safety standards, ban hazardous products, investigate safety of products, and use other forms of corrective action.
17. **State regulations.** State governments are increasingly active in the environmental and safety areas.

Source: U.S. Department of Commerce (1996).

DID YOU KNOW?

The costs of meeting mandated and self-imposed environmental and security requirements are large and continue to grow. Indeed, more than one-sixth of new P&E investment is for environmental/security improvement purposes rather than to improve productivity or increase output.

HOMELAND SECURITY DIRECTIVES

The rules, regulations, and standards listed in sidebar 3.3 were designed and implemented with the primary purpose of protecting workers on the job (OSHA standards) and to protect those residing outside the facility fenceline (USEPA rules, regulations, and programs). As a result of 9/11, the Homeland Security Department was formed. On matters pertaining to homeland security, Homeland Security Presidential Directives are issued by the president. Each directive has specific meaning and purpose and is carried out by the U.S. Department of Homeland Security. Each directive is listed and summarized in sidebar 3.4.

Sidebar 3.4. Homeland Security Presidential Directives

HSPD – 1: Organization and Operation of the Homeland Security Council. Ensures coordination of all homeland security–related activities among executive departments and agencies and promotes the effective development and implementation of all homeland security policies.

HSPD – 2: Combating Terrorism through Immigration Policies. Provides for the creation of a task force that will work aggressively to prevent aliens who engage in or support terrorist activity from entering the United States and to detain, prosecute, or deport any such aliens who are within the United States.

HSPD – 3: Homeland Security Advisory System. Establishes a comprehensive and effective means to disseminate information regarding the risk of terrorist acts to federal, state, and local authorities and to the American people.

HSPD – 4: National Strategy to Combat Weapons of Mass Destruction. Applies new technologies, increased emphasis on intelligence collection and analysis, strengthens alliance relationships, and establishes new partnerships with former adversaries to counter this threat in all of its dimensions.

HSPD – 5: Management of Domestic Incidents. Enhances the ability of the United States to manage domestic incidents by establishing a single, comprehensive national incident management system.

HSPD – 6: Integration and Use of Screening information. Provides for the establishment of the Terrorist Threat Integration Center.

HSPD – 7: Critical Infrastructure Identification, Prioritization, and Protection. Establishes a national policy for federal departments and agencies to identify and prioritize critical U.S. infrastructure and key resources and to protect them from terrorist attacks.

HSPD – 8: National Preparedness. Identifies steps for improved coordination in response to incidents. This directive describes the way federal departments and agencies will prepare for such a response, including prevention activities during the early stages of a terrorism incident. This directive is a companion to HSPD-5.

HSPD – 8 Annex 1: National Planning. Further enhances the preparedness of the United States by formally establishing a standard and comprehensive approach to national planning.

HSPD – 9: Defense of United States Agriculture and Food. Establishes a national policy to defend the agriculture and food system against terrorist attacks, major disasters, and other emergencies.

HSPD – 10: Biodefense for the Twenty-first Century. Provides a comprehensive framework for our nation's biodefense.

HSPD – 11: Comprehensive Terrorist-Related Screening Procedures. Implements a coordinated and comprehensive approach to terrorist-related screening that supports homeland security, at home and abroad. This directive builds upon HSPD – 6.

HSPD – 12: Policy for a Common Identification Standard for Federal Employees and Contractors. Establishes a mandatory, government-wide standard for secure and reliable forms of identification issued by the federal government to its employees and contractors (including contractor employees).

HSPD – 13: Maritime Security Policy. Establishes policy guidelines to enhance national and homeland security by protecting U.S. maritime interests.

HSPD – 15: U.S. Strategy and Policy in the War on Terror.

HSPD – 16: Aviation Strategy. Details a strategic vision for aviation security while recognizing ongoing efforts, and directs the production of a National Strategy for Aviation Security and supporting plans.

HSPD – 17: Nuclear Materials Information Program.

HSPD – 18: Medical Countermeasures against Weapons of Mass Destruction. Establishes policy guidelines to draw upon the considerable potential of the scientific community in the public and private sectors to address medical countermeasure requirements relating to CBRN threats.

(*continued*)

HSPD – 19: Combating Terrorist Use of Explosives in the United States. Establishes a national policy, and calls for the development of a national strategy and implementation plan, on the prevention and detection of, protection against, and response to terrorist use of explosives in the United States.

HSPD – 20: National Continuity Policy. Establishes a comprehensive national policy on the continuity of federal government structures and operations and a single National Continuity Coordinator responsible for coordinating the development and implementation of federal continuity policies.

HSPD – 21: Public Health and Medical Preparedness. Establishes a national strategy that will enable a level of public health and medical preparedness sufficient to address a range of possible disasters.

HSPD – 23: National Cyber Security Initiative.

HSPD – 24: Biometrics for Identification and Screening to Enhance National Security. Establishes a framework to ensure that federal executive departments use mutually compatible methods and procedures regarding biometric information of individuals, while respecting their information privacy and other legal rights.

Source: Department of Homeland Security (2008).

REFERENCES AND RECOMMENDED READING

Board on Chemical Sciences and Technology. 2006. *Terrorism and the chemical infrastructure: protecting people and reducing vulnerabilities.* Washington, DC: National Academies Press.

Bond, C. 2002. Statement on S. 2579. *Congressional Record,* Daily Edition, June 5, 1.

Brown, R. E., and B. R. Luce. 1990. *The value of pharmaceuticals: A study of selected conditions to measure the contribution of pharmaceuticals to health status.* Washington, DC: Battelle Medical Technology and Policy Research Center.

Bureau of Labor Statistics (BLS). 2008. *Career guide to industries, 2008–09 edition.* Chemical manufacturing, except pharmaceutical and medicine manufacturing. www.bls.gov/oco/cg/cgs008.htim (accessed February 16, 2008).

Department of Homeland Security (2008). Homeland Security Presidential Directives. www.dhs.gov/xabout/laws/editorial_0607.shtm (accessed February 6, 2009).

Dimes, J., et al. 1994. New drugs development in the United States from 1963–1964. *Clinical Pharmacology and Therapeutics* 55: 609–22.

Parachini, J. V. 2000. The World Trade Center bombers (1993). In *Toxic terror: Assessing terrorist use of chemical and biological weapons,* ed. J. B. Tucker, 185–206. Cambridge, MA: MIT Press.

Redwood, H. 1993. *Price regulation and pharmaceutical research.* Suffolk, England: Oldwicks Press.

U.S. Department of Commerce (USDC). 1996. *The chemical industry.* Washington, DC: U.S. Department of Commerce Office of Technology Policy.

U.S. International Trade Commission. 1991. Global competitiveness of U.S. advanced-technology manufacturing industries: Pharmaceuticals. Report to the Committee of Finance, U.S. Senate, on Investigation No. 332-302 Under Section 332(g) of the Tariff Act of 1930.

4

Chemical Facility Security

The emergence of amorphous and largely unknown terrorist individuals and groups operating independently (freelancers) and the new recruitment patterns of some groups, such as recruiting suicide commandos, female and child terrorists, and scientists capable of developing weapons of mass destruction, provide a measure of urgency to increasing our understanding of the psychological and sociological dynamics of terrorist groups and individuals. (Hudson 1999)

Note: At the time of this writing, the chemical facility security information from the Congressional Research Service, which is presented here, was last updated in January 2008.

First considered an environmental issue (e.g., hazardous materials spills contaminating the environmental media—air, water, and/or soil), chemical facility safety has been an issue of congressional interest for many years. Review of the historical incidents listed in sidebar 4.1 provides some perspective on the magnitude of the consequences that might result from terrorist attacks. Obviously, these incidents (and others) increase concern over the potential for release of toxic chemicals and the associated potential health impacts on surrounding areas.

Sidebar 4.1. Toxic or Flammable Chemical Disaster Scenarios

Event 1: Explosion—Ammonium nitrate. Azote de France Factory, Toulouse, France (September 21, 2001). 30 killed (7 off-site), 800 hospitalized, 2,400 injured, shock wave of 3.4 on the Richter scale, 50-foot crater resulted; 500 homes uninhabitable and 85 schools or colleges damaged; chemical releases and structural damages at other facilities (Dechy et al. 2004).

(*continued*)

Event 2: Explosion—Plastics manufacturing. Phillips Petroleum, Pasadena, Texas (October 23, 1989). 23 fatalities, 130–300 injured; extensive facility damage (U.S. Fire Administration 1989).

Event 3: Toxic Release (hydrofluoric acid vapor cloud)—Refinery. Marathon Refinery, Texas City, Texas (October 30, 1987). 4,000 people evacuated and more than 1,000 treated for injuries (Health and Safety Executive [UK] 2008a).

Event 4: Explosion—Unknown chemicals. BP Refinery, Texas City, Texas (March 23, 2005). 15 people killed and more than 100 wounded (CBS News 2006).

Event 5: Toxic Chemical Leak—Methyl isocyanate (MIC). Union Carbide Corporation, Bhopal, India (December 3, 1984). 3,000–7,000 people killed immediately; 20,000 cumulative deaths; 200,000–500,000 injured; post-traumatic stress; continued medical consequences (Lees 1996).

Event 6: Boiling Liquid Expanding Vapor Explosions (BLEVE)—Liquefied petroleum gas (LPG). PEMEX LPG Terminal, Mexico City (1984). 650 dead; 6,400 injured (Health and Safety Executive [UK] 2008b).

Event 7: Moored Ship Fire/Explosion—Ammonium nitrate. SS *Grandcamp*, Texas City, Texas (April 16, 1947). More than 560 killed and 2,000 injured (Texas City Firefighters Local 1259 2008).

Event 8: Chemical Spill—Oil. Ashland Oil Company, Inc., Floreffe, Pennsylvania (January 1988). The oil spill temporarily contaminated drinking water sources for an estimated 1 million people in Pennsylvania, West Virginia, and Ohio; contaminated river ecosystems; killed wildlife; damaged private property; and adversely affected businesses in the area. More than 511,000 gallons of diesel fuel remain unrecovered and are presumed to be in the rivers (U.S. Environmental Protection Agency 2008).

Event 9: Fire/Explosion—Sulfur dioxide and sulfuric acid. Motiva Enterprises, LLC, Delaware City, Delaware (July 17, 2001). Collapse of a spent sulfuric acid storage tank (more than 250,000 gallons) triggered releases from nearby tanks, killed one contract worker, and caused a large fish kill. Other commonly bermed tanks were immersed in concentrated sulfuric acid for several days until they could be drained, but they did not fail ("News Link" Environmental 2008).

Event 10: Explosion—Mononitrotoluene. First Chemical Corporation, Pascagoula, Mississippi (October 13, 2002). Three workers injured; fires, projectiles, and other damage to the plant and plant equipment (HighBeam 2008).

REGULATION OF CHEMICAL FACILITIES FOR SECURITY

Along with direction provided by the Homeland Security Presidential Directives listed in sidebar 4.1, statutory authority to the Department of Homeland Security (DHS) to regulate chemical facilities for security is provided by the Homeland Security Appropriations Act, 2007 (P.L. 109-295), Section 550. This law directed the Secretary of

Homeland Security to issue interim final regulations establishing risk-based performance standards for chemical facility security and requiring the development of vulnerability assessments and the development and implementation of site security plans. The law specified that these regulations are to apply only to those chemical facilities that the secretary determines present high levels of security risk. Consequently, some facilities are exempt from these regulations. Exempt facilities include those defined as a water system or wastewater treatment works; facilities owned or operated by the Department of Defense or Department of Energy; facilities regulated by the Nuclear Regulatory Commission; and those facilities regulated under the Maritime Transportation Security Act of 2002. (This act raises seaport security standards: P.L. 107-295.) Furthermore, the regulations are to allow regulated entities to employ combinations of security measures to meet the risk-based performance standards.

DID YOU KNOW?

A performance standard states requirements in terms of required results with criteria for verifying compliance but without stating the methods of achieving required results. A performance standard may define the functional requirements for the item, operational requirements, and/or interface and interchangeability characteristics. A performance standard may be viewed in juxtaposition to a prescriptive standard, which may specify design requirements, such as materials to be used, how a requirement is to be achieved, or how an item is to be fabricated or constructed (Office of Management and Budget 1998).

The secretary, under the law, must review and approve the required assessment, plan, and implementation for each facility. The statute prohibits the secretary from disapproving a site security plan on the basis of the presence or absence of a particular security measure, but the secretary may disapprove a site security plan that does not meet the risk-based performance standards. The secretary may approve vulnerability assessments (see chapter 5) and site security plans created through security programs not developed by DHS, so long as the results of these programs meet the risk-based performance standards established in regulation.

Note that the information developed for these regulations is to be protected from public disclosure but may be shared, at the secretary's discretion, with state and local government officials, including law enforcement officials and first responders possessing the necessary security clearances. Such shared information may not be publicly disclosed, regardless of state or local laws, and is exempt from the Freedom of Information Act (FOIA). Additionally, the information provided to the secretary, along with related vulnerability information, is to be treated as classified information in all judicial and administrative proceedings. Violation of the information protection provision is punishable by fine (Shea 2008).

The secretary not only can but must inspect and audit chemical facilities and determine regulatory compliance. If the secretary finds a facility not in compliance, the secretary must write to the facility explaining the deficiencies found, provide an opportunity for the facility to consult with the secretary, and issue an order to comply by a date determined by the secretary. If the facility continues to be out of compliance, the secretary may fine and, eventually, order the facility to close.

To enforce provisions of the law, only the Secretary of Homeland Security may bring a lawsuit against a facility owner. Note that the law does not affect any other federal law regulating chemicals in commerce. Interestingly, under the statute, states have the right to promulgate chemical facility security regulation that is at least as stringent as the federal chemical facility security regulation. Only in the case of an "actual conflict" (not defined) between the federal and state regulations would the state regulation be preempted. The federal statute contains a "sunset provision" and expires in 2009.

DID YOU KNOW?

The Secretary of Homeland Security determines which chemical facilities must meet the security requirements of DHS regulations based on the degree of risk posed by each facility. However, initially, screening of chemical facilities for compliance requirements was done on the basis of potential consequence rather than risk.

REFERENCES AND RECOMMENDED READING

CBS News. 2006. The explosion at Texas City. October 26. www.cbsnews.com/stories/2006/10/26/60minutes/main2126509.shtml (accessed April 27, 2008).

Crayton, J. W. 1983. Terrorism and the psychology of the self. In *Perspectives on terrorism*, ed. L. Z. Freedman and Y. Alexander, 33–41. Wilmington, DE: Scholarly Resources.

Dechy, N., T. Bourdeaux, N. Ayrault, M. A. Kordek, and J. C. Le Coze. 2004. First lessons of the Toulouse ammonium nitrate disaster, 21 September 2001, AZF plant, France. *Journal of Hazardous Materials* 111:131–38.

Erikson, E. 1994. *Identity of the life cycle.* New York: Norton.

Ferracuti, F. 1982. A sociopsychiatric interpretation of terrorism. *Annals of the American Academy of Political and Social Science,* 463 (1): 129–40.

Fields, R. M. 1979. Child terror victims and adult terrorists. *Journal of Psychohistory* 7 (1): 71–76.

Gurr, T. R. 1971. *Why men rebel.* Princeton, NJ: Princeton University Press.

Health and Safety Executive [UK] (HSE). 2008a. Release of hydrofluoric acid from Marathon Petroleum Refinery, Texas, USA. www.hse.gov.uk/comah/sragtech/casemarathon87.htm (accessed April 27, 2008).

———. 2008b. PEMEX LPG Terminal, Mexico City, Mexico. @www.hse.gov.uk/comah/sragtech/casepemex8484.htm (accessed April 27, 2008).

HighBeam. 2008. Explosion at First Chemical Corporation plant. www.highbeam.com/doc/1G1-92783471.html (accessed April 28, 2008).

Hudson, R. A. 1999. *The sociology and psychology of terrorism: Who becomes a terrorist and why?* Washington, DC: Library of Congress, Federal Research Division.

Lees, Frank. 1996. *Loss prevention in the process industries.* New York: Butterworth-Heinemann, 3:A5.1–A5.11.

Long, D. E. 1990. *The anatomy of terrorism.* New York: Free Press.

Margolin, J. 1977. Psychological perspectives on terrorism. In *Terrorism: Interdisciplinary perspectives,* ed. Y. Alexander and S. M. Finger. New York: John Jay Press.

"News Link" Environmental. 2008. Motiva Enterprises settles suit resulting from explosion at Delaware City refinery. www.caprep.com/0905038.htm (accessed April 27, 2008).

Office of Management and Budget (OMB). 1998. *Federal Conformity Assessment Activities, Circular No. A-119.* Washington, DC: Office of Management and Budget.

Olson, M. 1971. *The logic of collective action.* Boston: Harvard University Press.

Pearlstein, R. 1991. *The mind of the political terrorist.* Wilmington, DE: Scholarly Resources.

Shea, D. A. 2008. *Chemical facility security: Regulation and issues for Congress.* Washington, DC: Congressional Research Service.

Texas City Firefighters Local 1259. 2008. The Texas City disaster. www.local1259iaff.org/disaster.html (accessed April 27, 2008).

U.S. Environmental Protection Agency (USEPA). 2008. Ashland oil spill. www.epa.gov/reg3/PA/ashlandoil/.

U.S. Fire Administration (USFA). 1989. Phillips Petroleum Chemical Plant explosion and fire, Pasadena, Texas. U.S. Fire Administration Technical Report 035. Emmitsburg, MD: Federal Emergency Management Agency.

Wilkinson, P. 1974. *Political terrorism.* London: Macmillan.

5

Vulnerability Assessments

> Vulnerability means different things to different people. . . . Many associate vulnerability with a specific set of human activities. (Foster 1987)

One consequence of the events of 9/11 was the Department of Homeland Security (DHS) directive to establish a Critical Infrastructure Protection Task Force to ensure that activities to protect and secure vital infrastructure are comprehensive and carried out expeditiously. Another consequence is a heightened concern among citizens in the United States over the security of the chemical manufacturing and processing industry infrastructure. As mentioned, the nation's chemical industry infrastructure, consisting of more than sixty-six thousand separate operating facilities, is one of America's most valuable resources. Along with other critical infrastructure, the chemical industry is classified as "vulnerable," in the sense that inherent weaknesses in its operating environment could be exploited to cause harm to the system. There is also the possibility of a cascading effect—a chain of events—due to a terrorist act affecting a chemical facility that causes corresponding damage (collateral damage) to other nearby facilities. In addition to significant damage to the nation's chemical facilities—entities using raw chemical products to produce finished products—or other collateral damage, terrorist attacks against our chemical infrastructure would result in loss of life; catastrophic environmental damage to rivers, lakes, and wetlands; contamination of drinking-water supplies; long-term public health impacts; destruction of fish and shellfish production; and disruption to commerce, the economy, and our normal way of life.

VULNERABILITIES

Primarily because of the special dangers that flammable and toxic substances can pose, including immediate explosion and fire, release into the air or water, or theft and subsequent use in attacks elsewhere, ensuring the security of the chemical industry is important (Schierow 2004). As mentioned, the major groups of chemical products include basic industrial chemicals, plastics and rubbers, drugs, detergents, paints, and agricultural chemicals. However, only a small subset of the many thousands of chemicals produced and consumed in the United States are of concern for homeland security.

According to a 2004 Congressional Budget Office (CBO) report prepared for Congress, the USEPA—the primary agency tasked with protecting the public and the environment from chemical accidents—lists more than three hundred chemicals as "extremely hazardous." Focusing on the chemicals that could harm people after exposure for only a short time, the agency closely monitors chemical facilities with the capacity to process amounts in excess of threshold quantities from a list of seventy-seven acutely toxic chemicals and sixty-three flammable gases and liquids (Belke 2001).

DID YOU KNOW?

Threshold quantities for toxic substances range from 500 pounds to 20,000 pounds. For all listed flammable substances, the threshold quantity is 10,000 pounds. For explosive substances the threshold quantity is 5,000 pounds.

The most hazardous chemical substances come largely from three segments of the industry:

- petrochemicals (organic industrial chemicals)
- nitrates (agricultural chemicals)
- ammonia and chlorine (inorganic industrial chemicals)

Petrochemicals—such as fuels, solvents, and raw materials to make plastics—are of concern not only because they are highly flammable but also because they are found at very large and complex facilities, often in close proximity to one another. Thus, the immediate dangers from an attack and the resulting economic disruption would be greater than otherwise. In sufficient concentrations, the atmospheric release of many of those substances (by fire or spills) could also be immediately toxic or, over time, could be associated with cancer and other health problems or with environmental damage (CBO 2004).

Nitrates, produced from ammonia or urea, include a variety of nitrogen-based compounds used to make fertilizers, pesticides, and explosives, and they are highly flammable. Nitrate-fertilizer plants themselves could be targets for attack, but the fertilizers stored at thousands of warehouses across the country (accounting for the largest number of sites that the USEPA monitors) are perhaps of greater concern for homeland security, primarily because of the difficulty in securing fertilizers against misuse. The sale of ammonium nitrate fertilizer is legal. However, retail outlets could be vulnerable as targets for diversion of fertilizers by terrorists, and weakly guarded storage facilities could be targets for theft. Ammonium nitrate combined with diesel oil was used as the explosive in the World Trade Center bombing in 1993 and the Oklahoma City bombing in 1995 (CBO 2004).

Toxic substances such as chlorine and ammonia are a homeland security concern in part because of their widespread use in industrial applications and, hence, their accessibility. When released into the air or water in high concentrations, they can be very poisonous. The major use of ammonia and its compounds is as fertilizers.

Chlorine is also widely used in industrial applications. In water/wastewater treatment, for example, it is used in the disinfection unit process. Large quantities of chlorine gas (stored in one-ton cylinders or railroad tank cars) used in these treatment processes is a major concern of those tasked with ensuring plant security.

VULNERABILITY ASSESSMENT: THE PROCESS

For the purpose of this text and according to FEMA (2008), vulnerability is defined as any weakness that can be exploited by an aggressor to make an asset susceptible to hazard damage. A vulnerability assessment (VA) is an in-depth analysis of the facility's functions, systems, and site characteristics to identify facility weaknesses and lack of redundancy, and to determine mitigations or corrective actions that can be designed or implemented to reduce the vulnerabilities. During this assessment, the analysis of site assets is based on (1) the identified threat, (2) the criticality of the assets, and (3) the level of protection chosen (i.e., based on willingness or unwillingness to accept risk).

It is important to point out that post-9/11 the chemical industry sector has taken great strides to protect its critical infrastructure. For instance, government and industry have developed vulnerability assessment methodologies for critical infrastructure systems and trained thousands of auditors and others to conduct them.

Actually, before the Oklahoma City bombing, the anthrax attacks, and 9/11, OSHA and USEPA had already taken the initial steps to ensure safety and security in chemical production facilities and in other facilities that use, produce, or store listed chemicals. These steps were listed in OSHA's Process Safety Management (PSM) Standard and USEPA's Risk Management Program (RMP). Based on personal experience, having conducted VAs and modified VAs (wastewater) on both water and wastewater systems and having implemented RMP in a major U.S. wastewater system, we have to concur that these initial safety/security steps were quite effective.

The actual complexity of VAs will vary, based upon the design and operation of the chemical system. The nature and extent of the VA will differ among systems based on a number of factors, including system size and potential population. Safety evaluations also vary based on knowledge and types of threats, available security technologies, and applicable local, state, and federal regulations. Preferably, a VA is "performance-based," meaning that it evaluates the risk to the chemical industrial facility based on the effectiveness (performance) of existing and planned measures to counteract adversarial actions. According to USEPA (2002), the common elements of chemical industry vulnerability assessments are as follows:

- characterization of the chemical industrial system, including its mission and objectives
- identification and prioritization of adverse consequences to avoid
- determination of critical assets that might be subject to malevolent acts that could result in undesired consequences
- assessment of the likelihood (qualitative probability) of such malevolent acts from adversaries
- evaluation of existing countermeasures
- analysis of current risk and development of a prioritized plan for risk reduction

Vulnerability Assessment Tools

Several VA tools are available. One of the most commonly used tools is the Vulnerability Self-Assessment Tool (VSAT).

Chemical ISAC: Information Sharing

The Chemical Transportation Emergency Center (CHEMTREC®) created the Chemical Sector Information Sharing and Analysis Center (CHEM-ISAC). The CHEM-ISAC serves three purposes:

- to disseminate early warnings and alerts concerning threats against the physical infrastructure and cybersystems of chemical industry facilities
- to allow chemical industry facilities to share with each other information on security incidents
- to provide an opportunity for chemical facilities to have security incidents analyzed by counterterrorism experts

The types of information provided include:

- potential and imminent threats to chemical facilities
- reports of incidents nationwide

- incident trends
- possible responses to threats and attacks
- research on infrastructure protection

Information sources include:

- chemical facilities
- counterterrorism experts
- federal law enforcement agencies
- federal intelligence agencies
- USDOT
- industry research

Points to Consider in a Vulnerability Assessment (USEPA, 2002)

Some points to consider related to the six basic elements listed above are included in table 5.1. The manner in which the vulnerability assessment is performed is determined by each individual water/wastewater utility. Throughout the assessment process it is important to remember that the ultimate goal is twofold: to safeguard public health and safety and to reduce the potential for disruption of a reliable supply of chemicals.

Table 5.1. Basic elements in vulnerability assessments

Element	*Points to Consider*
1. Characterization of the chemical facility, including its mission and objectives	What are the important missions of the system to be assessed? Define the highest priority services provided by the utility. Identify the industry's customers: • general public • government • military • industrial • critical care • retail operations • firefighting
	What are the most important facilities, processes, and assets of the system for achieving the mission objectives and avoiding undesired consequences? Describe the: • industry facilities • operating procedures • management practices that are necessary to achieve the mission objectives • how the industry operates • treatment processes • storage methods and capacity • chemical use and storage • distribution system

(continued)

Table 5.1. *(continued)*

Element	*Points to Consider*
	In assessing those assets that are critical, consider critical customers, dependence on other infrastructures (e.g., electricity, transportation, communications), contractual obligations, single points of failure, chemical hazards and other aspects of the industry's operations, or availability of industry utilities that may increase or decrease the criticality of specific facilities, processes, and assets.
2. Identification and prioritization of adverse consequences to avoid	Take into account the impacts that could substantially disrupt the ability of the system to provide a safe and reliable supply of chemicals. Chemical facilities systems should use the vulnerability assessment process to determine how to reduce risk associated with the consequences of significant concern. Ranges of consequences or impacts for each of these events should be identified and defined. Factors to be considered in assessing the consequences may include • magnitude of service disruption • economic impact (such as replacement and installation costs for damaged critical assets or loss of revenue due to service outage) • number of illnesses or deaths resulting from an event • impact on public confidence in the chemical supply • chronic problems arising from specific events • other indicators of the impact of each event as determined by the chemical facility Risk reduction recommendations at the conclusion of the vulnerability assessment strive to prevent or reduce each of these consequences.
3. Determination of critical assets that might be subject to malevolent acts that could result in undesired consequences	What are the malevolent acts that could reasonably cause undesired consequences on these assets? • electronic, computer, or other automated systems that are utilized by the public water system (e.g., Supervisory Control and Data Acquisition (SCADA)) • the use, storage, or handling of various chemicals • the operation and maintenance of such systems
4. Assessment of the likelihood (qualitative probability) of such malevolent acts from adversaries (e.g., terrorists, vandals)	Determine the possible modes of attack that might result in consequences of significant concern based on critical assets of the system. The objective of this step of the assessment is to move beyond what is merely possible and determine the likelihood of a particular attack scenario. This is a very difficult task, as there is often insufficient information to determine the likelihood of a particular event with any degree of certainty. The threats (the kind of adversary and the mode of attack) selected for consideration during a vulnerability assessment will dictate, to a great extent, the risk reduction measures that should be designed to counter the threats. Some vulnerability assessment methodologies refer to this as a "design basis threat"

Element	*Points to Consider*
	(DBT), where the threat serves as the basis for the design of countermeasures as well as the benchmark against which vulnerabilities are assessed. It should be noted that there is no single DBT or threat profile for all chemical systems in the United States. Differences in geographic location, size of the utility, previous attacks in the local area, and many other factors will influence the threats that the chemical industry should consider in their assessments. Chemical industries should consult with the local FBI and/or other law enforcement agencies, public officials, and others to determine the threats upon which their risk reduction measures should be based.
5. **Evaluation of existing countermeasures** (Depending on countermeasures already in place, some critical assets may already be sufficiently protected. This step will aid in identification of the areas of greatest concern and help to focus priorities for risk reduction.)	What capabilities does the system currently employ for detection, delay, and response? • Identify and evaluate current detection capabilities such as intrusion detection systems, chemical quality monitoring, operational alarms, guard post orders, and employee security awareness programs. • Identify current delay mechanisms such as locks and key control, fencing, structural integrity of critical assets, and vehicle access checkpoints. • Identify existing policies and procedures for evaluation and response to intrusion and system malfunction alarms, and cybersystem intrusions. **It is important to determine the performance characteristics. Poorly operated and maintained security technologies provide little or no protection.** What cyber protection system features does the facility have in place? Assess what protective measures are in place for the SCADA and business-related computer information systems, such as • firewalls • modem access • Internet and other external connections, including wireless data and voice communications • security policies and protocols **It is important to identify whether vendors have access rights and/or "back doors" to conduct system diagnostics remotely.** What security policies and procedures exist, and what is the compliance record for them? Identify existing policies and procedures concerning: • personal security • physical security • key and access badge control • control of system configuration and operational data • chemical and other vendor deliveries • security training and exercise records

(continued)

Table 5.1. *(continued)*

Element	*Points to Consider*
6. Analysis of current risk and development of a prioritized plan for risk reduction	Information gathered on threats, critical assets, chemical production operations, consequences, and existing countermeasures should be analyzed to determine the current level of risk. The utility should then determine whether current risks are acceptable or risk reduction measures should be pursued. Recommended actions should measurably reduce risks by reducing vulnerabilities and/or consequences through improved deterrence, delay, detection, and/or response capabilities or by improving operational policies or procedures. Selection of specific risk reduction actions should be completed prior to considering the cost of the recommended action(s). Facilities should carefully consider both short- and long-term solutions. An analysis of the cost of short- and long-term risk reduction actions may impact which actions the utility chooses to achieve its security goals. Facilities may also want to consider security improvements. Security and general infrastructure may provide significant multiple benefits. For example, improved treatment processes or system redundancies can both reduce vulnerabilities and enhance day-to-day operation. Generally, strategies for reducing vulnerabilities fall into three broad categories: • Sound business practices, which affect policies, procedures, and training to improve the overall security-related culture at the chemical facility. For example, it is important to ensure that rapid communication capabilities exist between public health authorities and local law enforcement and emergency responders. • System upgrades, which include changes in operations, equipment, processes, or infrastructure itself that make the system fundamentally safer. • Security upgrades, which improve capabilities for detection, delay, or response.

RISK MANAGEMENT PROGRAM (RMP): CHEMICAL INDUSTRY

Whether hazardous chemicals are flammable or toxic, suppliers and users of those chemicals (including those who transport and store such materials) need to be (are required to be) concerned with the consequences of accidental and/or intentional release of dangerous chemicals into the environment. Based on data provided by businesses to USEPA's Risk Management Program (RMP), USEPA reported in 2000 that nearly fifteen thousand facilities were handling at least one hazardous substance in a quantity greater than threshold limits. Those facilities themselves represent a subset of a much larger number of businesses handling a "significant" quantity.

USEPA's Risk Management Program monitors large chemical producers, including petroleum refiners, petrochemical manufacturers, and nitrate-fertilizer manufactur-

Table 5.2. USEPA's number and percent of RMP-covered processes by industry sector

Industry Sector	*Number of Processes*	*Percent of Processes*
Agriculture and farming, farm supply, fertilizer production, pesticides	6,317	31
Water supply and wastewater treatment	3,753	18
Chemical manufacturing	3,803	18
Energy production, transmission, transport, and sale	3,038	15
Food and beverage manufacturing and storage (including refrigerated warehousing)	2,366	11
Chemical warehousing (not including refrigerated warehousing)	318	2
Pulp mills, iron and steel mills, cement manufacturing, computer manufacturing, etc.	1,075	5
Total	20,670	100

Source: GAO (2004).

ers. It also monitors many small facilities involved with fertilizer storage, refrigerated storage, and water treatment (see table 5.2). The greatest number of chemical processes that USEPA tracks involves just two toxic substances: ammonia and chlorine. Relatively few chemical processes, other than production of nitrate fertilizers, involve flammable substances. However, because the facilities that handle many flammable substances—petroleum refineries, petrochemical plants, natural gas plants, wholesale-fuel terminals, and propane distribution centers—have such large capacities, the greatest volume of hazardous chemicals that USEPA regulates are flammable.

In addition to the facilities monitored by the Risk Management Program, numerous facilities with smaller quantities of such chemicals can raise homeland security concerns. In particular, the fifteen thousand facilities in the program exclude retail outlets for flammable chemicals used as fuel, which are not required to report to USEPA. The many smaller suppliers, transporters, and consumers of those chemicals may hold sufficient quantities of dangerous materials to cause harm if the materials were released or set on fire. They may also have the information or equipment necessary to make even more dangerous substances. Further, because those smaller potential targets are especially dispersed and potentially more difficult to defend, they may be attractive as terrorist targets (GAO 2004).

Risk Management Program: Overview

> Community residents and industry officials do not consider the importance of accident prevention until after an accident occurs. . . . But then, the ghosts of Bhopal's victims must whisper, the only response can be: Too late. Too late. (Minter 1996, 6)

On May 24, 1996, USEPA finalized the Risk Management Program (RMP) under Section 112(r) of the 1990 Clean Air Act Amendments. On June 20, 1996, USEPA promulgated the new rule. The rule, under 40 CFR Part 68, is titled "Accidental Release Prevention Provisions: Risk Management Programs." Covered sources had until June 1999 to complete data, devise a risk management plan, institute a risk management program to comply with RMP, and submit the risk management plan to USEPA for review and approval (Spellman 1997).

Note: It should be pointed out that, as with OSHA's PSM, it is important to distinguish between a plan and a program. Specifically, the *plan* is the *information* and the document that the facility submits to the regulatory agency (USEPA for RMP) and maintains on-site for use by facility personnel. The *program*, however, is the *system* that backs up the plan and helps to ensure that the facility is operated according to the rule. The viable program is more than just a vehicle to be used in improving the facility's safety profile; it should also provide dividends as regards productivity, efficiency, and profitability. It is important to keep in mind that to be beneficial (i.e., to reduce accidents, injuries, and intrusions), the program, like any other management tool, must be upgraded and improved on a continuing basis. RMP is a living document.

As mentioned, RMP addresses specific chemicals/materials (compounds); it addresses the accidental release of over one hundred chemical substances. Of the RMP chemicals listed, seventy-seven include acutely toxic chemical compounds and sixty-three flammable gases. Threshold quantity levels range from 500 pounds to 20,000 pounds. USEPA estimates that approximately 100,000+ sources are covered by the rule. The universe includes chemical and most other manufacturers, certain wholesalers and retailers, drinking-water systems, wastewater treatment works, ammonia refrigeration systems, chemical wholesalers and end users, utilities, propane retailers, and federal facilities.

General Applicability

RMP includes three major parts or elements. These important elements are Hazard Assessment, Prevention Program, and Response Program; each of these is addressed in the following.

1. Hazard assessment. A hazard assessment is required to assess the potential effects of an accidental (or intentional) release of a covered chemical/material. This RMP element generally includes performing an off-site consequence analysis (OCA) and the compilation of a five-year accident history. The OCA must include analysis of a least one worst-case scenario. It must also include one alternative release scenario for the flammables class as a whole; also each covered toxic substance must have an alternative release scenario. USEPA has summarized some simplified consequence modeling approaches in an OCA guidance document. This OCA guidance document contains tables of dispersion and explosion modeling results that allow those who use them to minimize modeling efforts.

In its modeling requirement, USEPA has specified numerous mandatory modeling parameters and assumptions, primarily for the worst-case scenario analyses, to make OCAs more consistent. The worst-case scenario release quantity is defined as the largest vessel or pipe inventory, considering administrative controls that could limit the maximum inventory before the release. Generally, gas releases are assumed to occur over a ten-minute period; liquid pools are assumed to form instantaneously and then vaporize. Passive mitigation system credit may be given, if the system is capable of withstanding the release event of interest. For flammable releases, the analyst must assume that the entire release quantity vaporizes and undergoes a vapor cloud explosion.

Worst-case scenario. When considering the stationary source's worst-case scenario, there are selection factors to be considered. In addition to the largest inventories of a substance, the following conditions must also be considered: smaller quantities handled at higher process temperatures and pressures, and proximity to the boundary of the stationary source. Sources must analyze and report additional worst-case scenarios for a hazard class if the worst-case scenario from another covered process affects a different set of public receptors than the original worst-case scenario. It is interesting to note that worst-case release data indicate that the distances and thus the populations that could be threatened are greater for toxic substances than for flammable substances.

Alternative release scenario. In selecting alternative release scenarios, they must be more likely to occur than the worst-case scenario and must reach an endpoint off-site. In addition to those listed above, USEPA says owners should consider these factors in selecting alternative release scenarios: five-year accident history and failure scenarios identified by a process hazard analysis (PHA) or Program Level 2 hazard review. (Note: The three program levels are defined later in the text). The alternative release scenario analyses may be performed using somewhat more flexible modeling approaches and parameters than specified for worst-case scenario analyses. For example, active mitigation credit can be given.

Estimating distances. For both the worst-case and alternative release scenarios, the source must estimate the distance to where the endpoint is no longer exceeded and estimate the population (rounded to two significant digits) within a circle defined by the distance and centered at the release point. U.S. Census data may be used and it does not have to be updated; however, the presence of sensitive populations (e.g. hospitals, schools, etc.) must be noted. In addition, the source must identify and list the types of environmental receptors within the calculated worst-case distance and circle; however, no environmental damage assessment is required. In determining the presence of environmental receptors, U.S. Geological Survey maps may be used.

The off-site consequence analysis must be reviewed and updated every five years. However, if process changes might reasonably be expected to cause the worst-case scenario footprint or signature to increase or decrease by a factor of two or more, then the OCA must be revised and the risk management plan must be resubmitted to USEPA or designated authority within six months.

It is important to note that the five-year history must cover all accidental releases from covered processes that resulted in deaths, injuries, or significant property damage on-site, or known off-site deaths, injuries, evacuations, sheltering in place, property damage, or environmental damage. USEPA requires that ten specific types of accident data be compiled, including known initiating event, off-site impacts, contributing factors, and operation or process changes that resulted from investigation of the release.

2. Prevention program. A prevention program is required to prevent accidental releases of regulated substances. This element generally includes safety precautions and maintenance, monitoring, employee safety training, and other requirements similar to OSHA's PSM. It should be pointed out, however, that USEPA's requirements for the Program Level 2 elements (listed below) are less detailed than their OSHA PSM counterpart:

- safety information
- hazard reviews
- compliance audits
- maintenance
- operating procedures
- incident investigation
- training

For example, the hazard review requirements have the following differences from OSHA's process hazard analysis provision:

- no team requirement for the review
- fewer technical issues addressed in the analysis
- results to be documented and problem resolved in a timely manner; no requirement for a formal resolution system
- no requirement to keep all hazard review results for the life of the process
- no requirement to communicate finding to employees

Although the prevention program language of RMP's Program Level 2 is somewhat different from the requirements in the OSHA PSM Standard, this is not the case with the language of RMP's Program Level; it is virtually identical to that of the OSHA PSM Standard, except that the RMP rule uses different terms for some things (to be discussed later). These differences are based on the different legislative authorities that each agency holds. USEPA has also deleted specific phrases from the OSHA PSM regulatory language for the process safety information, process hazard analysis, and incident investigation elements to ensure that all sources implement process safety management in a way that protects not only workers, but also the public and the environment. Because of this language difference, companies should incorporate considerations of "off-site effects" into their OSHA process hazard analysis (PHA) revalidation protocols.

3. Response program. The response program requires specific action to be taken in emergency situations. This element generally includes procedures for notifying public and local agencies responsible for responding to accidental releases, information on emergency health care, and employee response training measures. These employee response training measures are required for plants whose employees are intended to respond to accidental releases using the plant's plan. The plan must address public notification, emergency medical treatment for accidental human exposures, and procedures of emergency response.

RMP Definitions and Requirements

The final management planning regulations (40 CFR Part 68) define the activities sources must undertake to address the risks posed by regulated substances in covered processes. To ensure that individual processes are subject to appropriate requirements that match their size and the risks they may pose, USEPA has classified them into three categories ("programs"). These program classifications are described in the following along with the requirements for regulated processes in each category.

Program 1. Requirements apply to processes for which a worst-case release, as evaluated in the hazard assessment, would not affect the public. These are sources or processes that have not had an accidental release that caused serious off-site consequences. Remotely located sources and processes using listed flammables are primarily those eligible for this program. Program 1 requirements are as follows:

1. Hazard assessment: worst-case analysis; five-year accident history.
2. Prevention program: Certify no additional steps needed.
3. Emergency response program: Coordinate with local responders.
4. Risk management plan contents: executive summary; registration; worst-case data; five-year accident history; certification.

Program 2. Requirements apply to less complex operations that do not involve chemical processing (e.g., retailers, propane users, non-chemical manufacturer's processes not regulated under OSHA's PSM standard). Program 2 requirements are as follows:

1. Hazard assessment: worst-case analysis; alternative releases; five-year accident history.
2. Management program: document management system.
3. Prevention program: safety information; hazard review; operating procedures; training; maintenance; incident investigation; compliance audit.
4. Emergency response program: Develop plan and program.
5. Risk management plan contents: executive summary; registration; worst-case data; alternative release data; five-year accident history; prevention program data; emergency response data; certification.

Program 3. Requirements apply to higher risk, complex chemical processing operations, to sources having a relevant process in one of the nine named SIC codes listed in table 5.3 or having process(es) subject to the OSHA PSM. Program 3 requirements are as follows:

1. Hazard assessment: worst-case analysis; alternative releases; five-year accident history.
2. Management program: document management system.
3. Prevention program: program safety information; process hazard analysis; operating procedures; training; mechanical integrity; incident investigation; compliance audit; management of change; pre-startup review; contractors; employee participation; hot work permits.
4. Emergency response program: Develop plan and program.
5. Risk management plan contents: executive summary; registration; worst-case data; alternative release data; five-year accident history; prevention program data; emergency response data; certification.

RMP/PSM: Similarities (Overlap)

It would be incorrect to compare RMP and PSM in the context of RMP versus PSM. The fact is that the RMP Rule and PSM Standard are designed to work together; they complement each other. This can be seen quite clearly when the similarities of the two regulations are illustrated.

For example, the OSHA PSM generally qualifies as meeting the RMP "prevention program" element. It is important to point out that in PSM, process safety techniques

Table 5.3. Program 3: SIC code applicability

SIC Code	*Industry*
2611	Pulp mills
2812	Alkalis and chlorine
2819	Industrial inorganics
2821	Plastics and resins
2865	Cyclic crudes
2869	Industrial organics
2873	Nitrogen fertilizers
2879	Agricultural chemicals
2911	Petroleum refineries

employ systematic methods for evaluating a process system and identifying potential hazards. For instance, such techniques as the checklist, "what if" analysis, fault tree analysis, event tree analysis, and hazard and operability studies (HAZOP) can be used. These techniques, used to conduct the process hazard analysis (PHA) for PSM, work well in satisfying the "prevention program" element requirement of RMP.

There are other complementary or similar elements shared by the two regulations. For example, both regulations share goals: (1) to prevent the accidental releases of regulated substances and (2) to minimize the consequences of releases that do occur.

At this point the reader is probably asking the obvious question: If RMP and PSM are so similar, then why do we need two different regulations? We need two different regulations because even though they are similar they are also different in a few ways. Before getting to these differences, however, there are still a few more similarities that need to be discussed.

Additional similarities (overlap) between RMP and PSM can be seen quite clearly if the facility under discussion or review is classified as a Program 3 facility. Assuming that a facility is categorized at the Program Level 3, then the following requirements will be familiar to those who have completed or followed the PSM requirements for their covered facility:

- process safety information
- process hazard analysis
- operating procedures
- training
- mechanical integrity
- management of change
- compliance audits
- incident investigation
- employee participation
- hot work permit

- contractors
- pre-startup review

It is interesting to note that facilities that are classified as Program 2 facilities do not have to include the management of change, pre-startup review, employee participation, hot work permit, or contractors' elements of PSM into their RMP.

RMP/PSM: Differences

The previous section discussed the similarities between RMP and PSM. Moreover, earlier it was pointed out that there are a few differences between the two regulations. In this section these differences will be discussed.

The first major difference between RMP and PSM is their source. RMP is a USEPA regulation. Along with its goal to reduce the harmful effects of accidental or intentional (terrorism) spills or releases, USEPA targets protection for those entities "outside the fence line." That is, USEPA is concerned with providing protection for the public; those who do not live or work in the covered facility. PSM, an OSHA regulation, on the other hand, targets its regulatory power toward ensuring protection to the workers—the personnel who work on the plant site. One could almost say that OSHA requires compliance inside the fence line only, as if it were a solitary, isolated entity. On the same note, one could also say that USEPA's RMP Rule knocks down this fence.

This difference in philosophy of who is to be protected by a particular regulation, the public or the worker, actually works to ensure that both are protected. This is the case because facilities affected by RMP generally are also affected by the requirements of PSM. Simply stated, by complying with the requirements of each regulation, both the pubic and the worker will be protected—and the environment benefits as well.

The USEPA's requirement to protect the public requires the covered facility to conduct an off-site consequence analysis (OCA). In PSM, the employer is only required to investigate each incident that resulted in or could have resulted in a catastrophic release of a highly hazardous chemical in the workplace.

There are other differences between RMP and PSM. This can be seen in the reporting requirements and in some of the different terms and definitions used by USEPA in RMP. To begin with, in regard to reporting requirements, under PSM OSHA requires the covered facility to comply with all applicable paragraphs. This compliance is expected to be completed by the covered facility but there is no reporting requirement (i.e., submission of a formal written document showing that compliance has been effected is not required under PSM).

However, this is not the case with RMP. In addition to requiring full compliance by those facilities covered under the regulation, it also requires each source to submit a risk management plan.

Table 5.4. PSM and RMP terms

PSM Term	*RMP Term*
Highly hazardous substance	Regulated substance
Employer	Owner or operator
Facility	Stationary source
Standard	Rules or part

With the exception of some key terms and phrases, the Program 3 prevention program language in RMP is identical to the OSHA PSM language. Most of the differences are in terminology based on specific legislative authorities given to USEPA or OSHA that have essentially the same meaning. To illustrate these differences, some of the RMP and PSM terms are listed in table 5.4.

In addition to using a few different key terms, RMP uses a few terms that are unique to the rule or are borrowed from PSM:

- *Off-site.* This is an area beyond the property boundary of the stationary source or an area within the property boundary to which the public has routine and unrestricted access during or outside business hours. Note: OSHA's jurisdiction includes visitors that may be on the property or facility who are conducting business as employees of other companies but does not necessarily extend to casual visitors or to areas within a facility boundary to which the public has routine and unrestricted access at any time.
- *Process.* USEPA has basically adopted OSHA's definition of process; USEPA decided to coordinate interpretations of the definition of process with OSHA to ensure that the rule is applied consistently.
- *Significant accidental release.* Any release of a regulated substance that has caused or has the potential to cause off-site consequences such as death, injury, or adverse effects to human health or the environment, or to cause the public to shelter in place or be evacuated to avoid such consequences.
- *Stationary source.* USEPA defined *source* to include the entire "facility." Sources are still required to submit one RMP and one registration as part of that RMP for all processes at the source with more than a threshold quantity of a regulated substance.

Summary of RMP Requirements

The owner or operator of a stationary source that has more than a threshold quantity (TQ) of a regulated substance in a process must

1. Prepare and submit a single risk management plan (RMP) that covers all affected processes and chemicals.

2. Program Level 1: Conduct a worst-case release scenario analysis, review accident history, ensure emergency response procedures are in place and coordinated with community officials.
3. Program Level 2: Conduct a hazard assessment, document a management system, implement a more extensive but still streamlined prevention program, and implement an emergency response program.
4. Program Level 3: Conduct a hazard assessment, document a management system, implement a prevention program that is basically identical to the OSHA PSM Standard, and implement an emergency response program.

RMP Hazard Review Checklists, What-If Questions, and HAZOP Procedures

The example checklist in sidebar 5.1 is designed for wastewater treatment plants using chlorine/sulfur dioxide.

Sidebar 5.1. Hazard Review Checklists

Checklist for Any WWTP

General Conditions, Operations, and Maintenance	*Yes / No / n/a*	*Comments*
Are work areas clean?		
Are adequate warning signs posted?		
Is ambient temperature normally comfortable?		
Is lighting sufficient for all operations?		
Are the right tools provided and used?		
Is personal protective equipment (PPE) provided and adequate?		
Are containers and tanks protected from vehicular traffic?		
Are all flammable and combustible materials kept away from containers, tanks, and feed lines?		
Are containers, tanks, and feed line areas kept free of any objects that can fall on them (e.g., ladders, shelves?)		
Are leak detectors with local and remote audible and visible alarms present, operable, and tested?		
Are windsocks provided in a visible location?		
Are emergency repair kits available for each type of supply present?		
Are appropriate emergency supplies and equipment present, including PPE and self-contained breathing apparatus (SCBA)?		
Are emergency numbers posted in an appropriate spot?		
Are equipment, containers, and railcars inspected daily?		
Are written operating procedures available to the operators?		
Are preventive maintenance, inspections, and testing performed as recommended by the manufacturer and industry groups and documented?		

Human Factors	*Yes / No / n/a*	*Comments*
Have operators been trained on the written operating procedures and the use of PPE in normal operations (or for operators on the job		

before June 21, 1999, have you certified that they have the required knowledge, skills, and ability to do their duties safely)?		
Do the operators follow the written operating procedures?		
Do the operators understand the applicable operating limits on temperature, pressure, flow, and level?		
Do the operators understand the consequences of deviations above or below applicable operating limits?		
Have operators been trained on the correct response to alarms and conditions that exceed the operating limits of the system?		
Are operators provided with enough information to diagnose alarms?		
Are controls accessible and easily understood?		
Are labels adequate on instruments and controls?		
Are all major components, valves, and piping clearly and unambiguously labeled?		
Are all components mentioned in the procedures adequately labeled?		
Are safe work practices, such as lockout/tagout, hot work, and line opening procedures followed?		
Are personnel trained in the emergency response plan and the use of emergency kits, PPE, and SCBAs?		
Are contractors used at the facility?		
Are contractors trained to work as safely as your own employees?		
Do you have programs to monitor that contractors are working safely?		

Chlorine and Sulfur Dioxide—Siting	*Yes / No / n/a*	*Comments*
Are material safety data sheets (MSDSs) readily available to those operating and maintaining the system?		
Do employees understand that there are certain materials with which chlorine/sulfur dioxide must not be mixed?		
Do employees understand the toxicity, mobility, and ability of chlorine/sulfur dioxide to sustain combustion?		
Do employees understand the consequences of confining liquid chlorine/sulfur dioxide without a thermal expansion device?		
Do employees understand the effect of moisture on the corrosive potential of chlorine/sulfur dioxide?		
Do employees understand the effects of fire and elevated temperature on the pressure of confined chlorine/sulfur dioxide and the potential for release?		

Chlorine and Sulfur Dioxide—Container Shipment Unloading	*Yes / No / n/a*	*Comments*
Is the truck inspected for wheel chocks, proper position, and condition of crane?		
Are adequate warning signs posted? Are there "stops"?		
Is the shipment inspected for leakage, general condition, currency of hydrostatic test, and valve protective housing before accepting?		
Are containers placed in the 6 o'clock/12 o'clock position for storage to reduce the chance of a liquid leak through the valve?		

Chlorine and Sulfur Dioxide—Bulk Shipment Unloading	*Yes / No / n/a*	*Comments*
Do procedures call for hand brakes to be set and wheels chocked before unloading?		

(conituned)

Do procedures call for safety systems to be inspected prior to making connections for unloading or between storage tanks and transfer or distribution systems?

For railcars, are derails to protect the open end located at least 50 feet from the car being protected?

Are railcars staged at dead-end tracks and guarded against damage from other railcars and motor vehicles?

Are caution signs placed at each derail and as appropriate in the vicinity of chlorine/sulfur dioxide storage, use, and transfer areas?

Does the transfer operation incorporate an emergency shutoff system?

Is a suitable operating platform provided at the transfer station for easy access and rapid escape?

Is padding air for railcars from a dedicated, flow-limited, dry (to −40°F or below), and oil-free source?

Is tank care attended to as long as the car is connected, in accordance with DOT regulations?

Building and Housing Chlorine/Sulfur Dioxide Systems	*Yes / No / n/a*	*Comments*
Does the building conform with local building and fire codes and NFPA-820?		
Is the building constructed of noncombustible materials?		
Is continuous leak detection, using area chlorine/sulfur dioxide monitors, provided in storage and process areas?		
If flammable materials are stored or used in the same building, are they separated from the chlorine/sulfur dioxide areas by a fire wall?		
Are two or more exits provided from each chlorine/sulfur dioxide storage and process area and building?		
Is the ventilation system appropriately designed for indoor operation (and scrubbing, if required) by local codes in effect at the time of construction or major modification?		
Are the exhaust ducts near floor level and the intake elevated?		
Can the exhaust fan be remotely started and stopped?		
If chlorine and sulfur dioxide are stored in the same building, are storage rooms separated as required?		

Chlorine and Sulfur Dioxide—Piping and Appurtenances	*Yes / No / n/a*	*Comments*
Do piping specifications meet chlorine/sulfur dioxide requirements for the service?		
Do you require suppliers to provide documentation that all piping and appurtenances are certified "for chlorine service" or "for sulfur dioxide service" by the manufacturer?		
Are piping systems properly supported, adequately sloped to allow drainage, and with a minimum of low spots?		
Is all piping protected from all risks of excessive fire or heat?		
Is an appropriate liquid expansion device or vapor pressure relief provided on every line segment or device that can be isolated?		

Chlorine and Sulfur Dioxide—Design Stage of New/Modified Process	*Yes / No / n/a*	*Comments*
Is the system designed to operate at lowest practical temperatures and pressures?		
If chlorine/sulfur dioxide is low enough, is the system designed to feed gaseous chlorine/sulfur dioxide from the storage container, rather than liquid?		

Have the lengths of liquid chlorine/sulfur dioxide lines been minimized (reduces quantity of chlorine in lines available for release)?

Are low-pressure alarms and automatic shutoff valves provided on chlorine/sulfur dioxide feed lines?

Are vent-controlled spill collection sumps provided and floors sloped toward sumps for stationary tanks and railcars?

Are vaporizers provided with automatic gas line shutoff valve, downstream pressure-reducing valve, gas flow control valve, temperature control system and interlocks to shut down gas flow on low vaporizer temperature, and appropriate alarms in a continuously manned control room?

Do vaporizers have a limited heat input capacity?

Are curbs, sumps, and diking that minimize the surface or potential spills provided for stationary tanks and railcars?

Checklist for Anhydrous Ammonia Systems

Anhydrous Ammonia—Basic Rules	*Yes / No / n/a*	*Comments*
Does the storage tank have a permanently attached nameplate?		
Are containers at least 50 feet from wells or other sources of potable water supply?		
Are containers painted white or other light-reflecting colors and maintained in good condition?		
Is the area free of readily ignitable materials?		
Are all main operating valves on tanks identified to show liquid or vapor service?		

Anhydrous Ammonia—Appurtenances	*Yes / No / n/a*	*Comments*
Are all appurtenances designed for maximum working pressure and suitable for ammonia service?		
Do all connections to containers have shut-off valves as close to container as practicable (except safety relief devices and gauging devices)?		
Are the excess flow and/or back pressure check valves located inside the container or at a point outside as close as practicable to where line enters container?		
Are excess flow valves plainly and permanently marked with name of manufacturer, catalog number, and rated capacity?		

Anhydrous Ammonia—Piping	*Yes / No / n/a*	*Comments*
Are piping and tubing suitable for ammonia service?		
Are provisions made for expansion, contraction, jarring, vibration, and settling?		
Is all exposed piping protected from physical damage from vehicles and other undue strain (2,000 lb. pull)?		

Anhydrous Ammonia—Hoses	*Yes / No / n/a*	*Comments*
Does the hose conform to TFI-RMA specifications for anhydrous ammonia?		
Is it 350 psig working, 1750 psig—burst?		

(continued)

Is it marked every 5 feet with "Anhydrous ammonia, ___ psig" (maximum working pressure), manufacturer's name or trademark, year of manufacture?

Anhydrous Ammonia—Safety Relief Devices	*Yes / No / n/a*	*Comments*
Are safety relief valves installed?		
Are they vented upward and unobstructed to the atmosphere?		
Do they have a rain/dust cap?		
Are shut-off valves not installed between safety relief valve and container?		
Are safety relief valves marked with "NH3" or "AA", psig valve set to start to discharge, CFM flow at full open, and manufacturer's name and catalog number?		
Is flow capacity restricted on upstream or downstream side?		
Are hydrostatic relief valves installed between each pair of valves in liquid piping or hose?		

Anhydrous Ammonia—Safety	*Yes / No / n/a*	*Comments*
Are there two suitable full face masks with ammonia canisters as approved by the Bureau of Mines? Is self-contained breathing air apparatus required in concentrated atmospheres?		
Is an easily accessible shower or a 50-gallon drum of water available?		

Anhydrous Ammonia—Transfer of Liquid	*Yes / No / n/a*	*Comments*
Are pumps designed for ammonia service and at least 250 psig working pressure?		
Does PD pump have relief valve installed?		
Is a 0–400 psi pressure gauge installed on pump discharge?		
Are loading/unloading lines fitted with backflow check or excess flow valves?		
Are caution signs posted when railcars are loading/unloading?		
Are containers equipped with an approved liquid level gauging device (except those filled by weight)?		
Are containers fitted with a fixed tube liquid level gauge at 85% of water capacity?		

Anhydrous Ammonia—Stationary Tank	*Yes / No / n/a*	*Comments*
Are nonrefrigerated containers designed for a minimum 250 (265 in CA) psig pressure?		
Are all liquid and vapor connections to containers, except safety relief valves, liquid gauging, and pressure gauge connections fitted with orifices not larger than No. 54 drill size, equipped with excess-flow valves?		
Are storage containers fitted with a 0–400 psi ammonia gauge?		
Are they equipped with vapor return valves?		
Are containers marked on at least two sides with "Anhydrous Ammonia" or "Caution—Ammonia" in contrasting colors and minimum 4-inch-high letters?		
Is a sign displayed stating name, address, and phone number of nearest representative, agent, or owner?		
Are containers installed on substantial concrete, masonry, or structural steel supports?		

Are ammonia systems protected from possible damage by moving vehicles?

Anhydrous Ammonia—Basic Rules	*Yes / No / n/a*	*Comments*
Are storage tanks painted white or other light-reflecting colors and maintained in good order?		
Is storage area free of readily ignitable materials?		
Are storage tanks kept away from wells or other sources of potable water supply?		
Are storage tanks located with ample working space all around?		
Are storage tanks properly vented and away from areas where operators are likely to be?		
Does receiving system include a vapor return?		
Is storage capacity adequate to receive full volume of delivery vehicle?		
Are storage tanks secured against overturn by wind, earthquake, and/or flotation?		
Are tank bottoms protected from external corrosion?		
Is aqua ammonia system protected from possible damage from moving vehicles?		
Are storage tanks labeled as to content?		
Are all appurtenances suitable for aqua ammonia service?		
Are all storage tanks fitted with liquid level gauges?		
Are liquid level gauges adequately protected from physical damage?		
If tubing is used, is it fitted with a fail-closed valve?		
Are all storage tanks fitted with overfill fittings or high-level alarms?		
Are tanks fitted with pressure/vacuum valves?		
Is an ammonia gas scrubber system used?		
Are piping and hose materials suitable for aqua ammonia service?		
Is piping free of stain and provision made for expansion, contraction, jarring, vibration, and settling?		
Is all exposed piping protected from physical damage from moving vehicles and other undue strain?		
Are hoses securely clamped to hose barbs?		
Are hoses inspected and renewed periodically to avoid breakage?		
Are pumps designed for aqua ammonia service?		
Are pumps fitted with splash guard around seals?		
Are pumps fitted with coupling guards?		
Do pumps have local start/stop stations?		
Are two suitable full face masks available with ammonia canisters as approved by the Bureau of Mines?		
Is a self-contained breathing air apparatus required in concentrated atmospheres?		
Is an easily accessible quick-acting shower with bubble fountain or 250-gallon drum of clean water available?		
Is an extra pair of chemical splash-proof goggles and/or full face shields available?		
Is an extra set of ammonia-resistant gloves, boots, coat, and apron available?		

(*conitunued*)

Are fire extinguishers and a first aid kit available?

Are handlers/operators wearing their goggles and gloves when working with aqua ammonia?

Is safety and first aid information posted?

Are emergency phone numbers and individuals to contact posted?

Source: Spellman (1997).

PSM/RMP Process Hazard Analysis (PHA) Methodologies

OSHA/USEPA requires employers, such as the chemical industry service sector, to perform an initial process hazard analysis (PHA) on processes covered by PSM/RMP standards. The PHA must be appropriate to the complexity of the process and must identify, evaluate, and control the hazards involved in the process. Employers are required to determine and document the priority order for conducting process hazard analyses based on a rationale that includes such considerations as extent of the process hazards, number of potentially affected employees, age of the process, and operating history of the process.

As mentioned earlier, OSHA's PSM is primarily concerned with protecting "employees" from the effects of accidental/intentional discharge of covered hazardous materials/chemicals—protecting those within the fence line, so to speak. USEPA's RMP, on the other hand, is concerned with protecting those residing outside the plant's fence line; this also includes protecting the environmental media: air, water, and soil. Notwithstanding the benefits to be gained from performance of vulnerability assessment (VA), in combination, PSM/RMP can enhance a facility's overall security profile.

The PHA procedure can be conducted using various methodologies. For example, the checklist analysis discussed earlier is an effective methodology. In addition, Pareto analysis, relative ranking, pre-removal risk assessment (PRRA), change analysis, failure mode and effects analysis (FMEA), fault tree analysis, event tree analysis, event and CF charting, PrHA, what-if analysis, and HAZOP can be used in conducting the PHA.

Based on personal experience, the what-if analysis and HAZOP seem to be the most user-friendly methodologies to use. In the following example we describe the what-if analysis procedure and sample checklists typically used for chemicals used in wastewater treatment. Moreover, guide words, meanings, process parameters, and procedures for conducting HAZOP are also presented.

What-If Analysis Procedure / Sample What-If Questions

The steps in a what-if checklist analysis are as follows:

1. Select the team (personnel experienced in the process).
2. Assemble information (piping and instrumentation drawings (P&IDs), process flow diagrams (PFDs), operating procedures, equipment drawings, etc.).

3. Develop a list of what-if questions.
4. Assemble your team in a room where each team member can view the information.
5. Ask each what-if question in turn and determine:
   - What can cause the deviation from design intent that is expressed by the question?
   - What adverse consequences might follow?
   - What are the existing design and procedural safeguards?
   - Are these safeguards adequate?
   - If these safeguards are not adequate, what additional safeguards does the team recommend?
6. As the discussion proceeds, record the answers to these questions in tabular format.
7. Do not restrict yourself to the list of questions that you developed before the project started. The team is free to ask additional questions at any time.
8. When you have finished the what-if questions, proceed to examine the checklist in sidebar 5.2. The purpose of this checklist is to ensure that the team has not forgotten anything. While you are reviewing the checklist, other what-if questions may occur to you.
9. Make sure that you follow up all recommendations and action items that arise from the hazards evaluation.

Sidebar 5.2. What-If Questions

What-If Questions for Chlorine and Sulfur Dioxide Systems

1. Movement of 1-ton chlorine/sulfur dioxide cylinders

   What if the cylinder is dropped from the lifting apparatus?

   What if the truck rolls forward or backward?

   What if a cylinder rolls and drops from the truck?

   What if the cylinder swings while being lifted?

   What if the chlorine/sulfur dioxide container is not empty when removed from service?

   What if the automatic container switchover system fails?

   What if a chlorine/sulfur dioxide cylinder is delivered instead of sulfur dioxide/chlorine?

   What if the cylinder is not in good condition?

2. Ton cylinders on trunnion, including pigtails, (subheader lines) to main header lines

   What if pigtails rupture while connected on-line?

   What if pigtail connections open or leak when pressure is applied?

(*continued*)

What if something is dropped onto cylinder or connection?

What if cracks develop in the ton cylinder flexible connection?

What if liquid chlorine/sulfur dioxide is withdrawn through the vapor lines from the ton cylinder?

What if the cylinder valve cannot be closed during an emergency?

What if there are pinholes or small leaks at the fusible plugs?

What if ton cylinder ends change shape from concave to convex?

What if liquid is trapped between two closed valves and the temperature rises?

What if there is a fire near the cylinders?

What if the operator leaves the valve open and disconnects the pigtail?

What if water enters the systems?

3. Chlorine/sulfur dioxide headers in the chlorination (sulfonation) room

What if the pressure relief valve sticks open?

What if a valve leaks?

What if there is inadequate flow in the gas line (e.g., filter clogged)?

4. Evaporators

What if there is overpressure in the evaporator?

What if there is low temperature in the evaporator?

What if rupture disks leak?

What if the vacuum regulator valve fails?

What if there is a gas pressure gauge leak?

What if the vacuum regulator check unit fails?

What if there is liquid chlorine/sulfur dioxide carryover to the vacuum regulating valve downstream of the evaporator?

5. Chlorination (sulfonation) and pipes to injectors

What if there are leaks in the chlorinator (sulfonator) unit?

What if there is rupture of the pipe from the chlorinator to the injector?

What if there is backflow of water into the chlorine/sulfur dioxide line?

What if the water pump is not working?

6. General

What if there is a power failure?

What if chlorine/sulfur dioxide is released during maintenance?

What if a chlorine/sulfur dioxide leak is not detected?

What if there is moisture in the chlorine/sulfur dioxide system?

7. Scrubbers

What if the system loses scrubber draft?

What if the system loses scrubber solution?

What if the manual vent to the scrubber is opened during operation?

What if the leak tightness of the building is compromised during emergency operation of the scrubbers?

8. Tank trucks

   What if the liquid hose leaks or ruptures?

   What if the vapor return hose leaks or ruptures?

   What if the truck moves?

   What if the mass of chlorine/sulfur dioxide in the truck exceeds the capacity of the tank?

   What if the chlorine tank truck is connected to a sulfur dioxide vessel (or vice versa)?

   What if there is something other than chlorine (or sulfur dioxide) in the truck?

   What if there is a fire under or near the truck?

   What if the truck collides with pipework or building housing chlorine/sulfur dioxide storage vessels?

9. Railcars

   What if the liquid hose leaks or ruptures?

   What if the padding air is moist?

   What if the padding air hose ruptures?

   What if the railcar moves?

   What if the relief valve lifts below the set pressure?

   What if there is a fire under or near the truck?

   What if there is a fire on or near the railcar?

What-If Questions for Ammonia Systems

1. Storage vessel

   What if the vessel is overfilled?

   What if there is fire under or near the vessel?

   What if the relief valve fails to lift on demand?

   What if the relief valve opens below its set pressure?

   What if the deluge system fails to work on demand?

2. Tank truck unloading

   What if the liquid unloading hose partially ruptures?

   What if the liquid unloading hose completely ruptures?

   What if the tank truck moves?

   What if the tank truck drives away before the hose is disconnected?

(*continued*)

What if the vapor return hose partially or completely ruptures?

What if valves are not completely closed before disconnecting the hoses?

What if the tank truck contains something other than ammonia?

What if the ammonia in the tank truck contains excess oxygen?

What-If Questions for Digester Systems

What if something falls onto a digester cover?

What if relief valves on a digester open?

What if an intermediate digester gas storage vessel fails?

What if air is introduced into the gas collection system?

What if the gas collection header leaks or ruptures or becomes blocked?

What if a digester gas compressor fails catastrophically?

What if there is a digester gas leak into a building (digester building, compressor room, boiler room)?

What if the digester gas pressure exceeds the cover pressure rating?

What if the floating digester gas cover jams or tilts?

General What-If Questions

What if the ambient temperature is abnormally high?

What if the ambient temperature is abnormally low?

What if there is a hurricane?

What if there is a tornado?

What if there is flooding?

What if there is a heavy snowfall?

What if there is an earthquake?

What if there is a tidal wave?

What if there is a failure of electric power?

HAZOP Analysis

The HAZOP analysis technique uses a systematic process to (1) identify possible deviations from normal operations and (2) ensure that safeguards are in place to help prevent accidents. The HAZOP uses special adjectives (such as speed, flow, pressure, etc.; see table 5.5) combined with process conditions (such as "more," "less," "no," etc.; see table 5.6) to systematically consider all credible deviations from normal conditions. The adjectives, called guide words, are a unique feature of HAZOP analysis.

In this approach, each guide word is combined with relevant process parameters and applied at each point (study node, process section, or operating step) in the process that is being examined (see table 5.7).

Guide words are applied to both the more general parameters (e.g., react, mix) and the more specific parameters (e.g., pressure, temperature). With the general param-

Table 5.5. Common HAZOP analysis process parameters

Flow	*Time*	*Frequency*	*Mixing*
Pressure	Composition	Viscosity	Addition
Temperature	pH	Voltage	Separation
Level	Speed	Information	Reaction

Table 5.6. Guide words and their meanings

Guide Words	*Meaning*
No	Negation of the design intent
Less	Quantitative decrease
More	Quantitative increase
Part of	Other materials present by intent
As well as	Other materials present unintentionally
Reverse	Logical opposite of the intent
Other than	Complete substitution

Table 5.7. Examples of creating deviations using guide words and process parameters

Guide Words		*Parameter*		*Deviation*
No	+	Flow	=	No Flow
More	+	Pressure	=	High Pressure
As well as	+	One phase	=	Two Phase
Other than	+	Operation	=	Maintenance
More	+	Level	=	High Level

eters, it is not unusual to have more than one deviation from the application of one guide word. For example, "more reaction" could mean either that a reaction takes place at a faster rate, or that a greater quantity of product results. On the other hand, some combination of guide words and parameters will yield no sensible deviation (e.g., "as well as" with "pressure").

HAZOP Procedure

1. Select the team.
2. Assemble information (P&IDs, PFDs, operating procedures, equipment drawings, etc.).
3. Assemble your team in a room where each team member can view P&IDs.
4. Divide the system you are reviewing into nodes (you can present the nodes, or the team can choose them as you go along).
5. Apply appropriate deviations to each node. For each deviation, address the following questions:
   - What can cause the deviation from design intent?
   - What adverse consequences might follow?
   - What are the existing design and procedural safeguards?

- Are these safeguards adequate?
- If these safeguards are not adequate, what does the team recommend?

6. As the discussion proceeds, record the answers to these questions in tabular format.

Final Word on PSM/RPM Compliance

Governor Tom Ridge points out the security role for public service professionals:

> Americans should find comfort in knowing that millions of their fellow citizens are working every day to ensure our security at every level—federal, state, county, municipal. These are dedicated professionals who are good at what they do. I've seen it up close, as governor of Pennsylvania. . . . But there may be gaps in the system. The job of the Office of Homeland Security will be to identify those gaps and work to close them. (Henry 2002)

It is to shore up the "gaps in the system" that has driven many chemical industrial facilities to increase security. In addition to vulnerability assessments (VA) and PSM/RMP provisions, other security steps should also be taken. For example, it is the opinion of the authors of this text that the following recommendations are the "minimum" steps to be taken to upgrade security. The recommendations include

1. **Guard against unplanned physical intrusion.**
   - Lock all doors and set alarms at your facility.
   - Limit access to facilities and control access to pumping stations, and to chemical and fuel storage areas, giving close scrutiny to visitors and contractors.
   - Post guards at chemical plants, and post "Employee Only" signs in restricted areas;
   - Secure hatches, metering vaults, manholes, and other access points to the facility.
   - Increase lighting in parking lots, treatment bays, and other areas with limited staffing.
   - Control access to computer networks and control systems, and change the passwords frequently.
   - Do not leave keys in equipment or vehicles at any time.
2. **Make security a priority for employees.**
   - Conduct background security checks on employees at hiring and periodically thereafter.
   - Develop a security program with written plans and train employees frequently.
   - Ensure all employees are aware of communications protocols with relevant law enforcement, public health, environmental protection, and emergency response organizations.

- Ensure that employees are fully aware of the importance of vigilance and the seriousness of breaches in security, and that they make note of unaccompanied strangers on the site and immediately notify designated security officers or local law enforcement agencies.
- Consider varying the timing of operational procedures if possible, in case someone is watching the pattern changes.
- Upon the dismissal of an employee, change passcodes and make sure keys and access cards are returned.
- Provide customer service staff with training and checklists about how to handle a threat if it is called in.

3. **Coordinate actions for effective emergency response.**
   - Review existing emergency response plans, and ensure they are current and relevant.
   - Make sure employees have necessary training in emergency operating procedures.
   - Develop clear protocols and chains-of-command for reporting and responding to threats along with relevant emergency, law enforcement, environmental, and public health officials, consumers, and the media. Practice the emergency protocols regularly.
   - Ensure key plant or facility personnel (both on and off duty) have access to crucial telephone numbers and contact information at all times. Keep the call list up to date.
   - Develop close relationships with local law enforcement agencies, and make sure they know where critical assets are located. Request they add your facilities to their routine rounds.
   - Report to county or state health officials any illness among the employees that might be associated with chemical contamination.
   - Report criminal threats, suspicious behavior, or attacks on wastewater utilities immediately to law enforcement officials and the relevant field office of the Federal Bureau of Investigation.
4. **Invest in security and infrastructure improvements.**
   - Assess the vulnerability of chemical and fuel storage areas, and other key infrastructure elements.
   - Move as quickly as possible with the most obvious and cost-effective physical improvements, such as perimeter fences, security lighting, tamper-proof manhole covers and valve boxes, etc.
   - Improve computer system and remote operational security.
   - Use local citizen watches.
   - Seek financing for more expensive and comprehensive system improvements.

THE BOTTOM LINE ON SECURITY

Again, when it comes to the security of chemical industry infrastructure (and the rest of our nation), few have summed it up better than Governor Ridge.

> Now, obviously, the further removed we get from September 11, I think the natural tendency is to let down our guard. Unfortunately, we cannot do that. The government will continue to do everything we can to find and stop those who seek to harm us. And I believe we owe it to the American people to remind them that they must be vigilant, as well. (Henry 2002)

REFERENCES AND RECOMMENDED READING

Belke, J. C. 2001. Chemical accident risks in U.S. industry: A preliminary analysis of accident risk data from U.S. hazardous chemical facilities. In *Proceedings of the Tenth International Symposium on Loss Prevention and Safety Promotion in the Process Industries, Stockholm, Sweden.* Amsterdam: Elsevier Science.

Clark, R. M., and R. A. Deininger. 2000. Protecting the nation's critical infrastructure: The vulnerability of U.S. water supply systems. *Journal of Contingencies and Crisis Management* 8 (2): 76–80.

Congressional Budget Office (CBO). 2004. Homeland Security and the private sector. www.cbo.gov/ftpdocs/60xx/doc6042/12-20-HomelandSecurity.pdf (accessed May 2, 2008).

Federal Emergency Management Agency (FEMA). 2008. FEMA 452—Risk assessment: A how-to guide to mitigate potential terrorist attacks. www.fema.gov/library/viewRecord.do?id=1938 (accessed May 1, 2008).

Foster, S. S. D. 1987. Fundamental concepts in aquifer vulnerability, pollution risk, and protection strategy. In *Vulnerability of soil and groundwater to pollutants,* ed. W. Van Duijvenbooden and H. G. van Waegeningh. The Hague: TNO Committee on Hydrological Research.

Government Accountability Office (GAO). 2004. *Homeland security: Federal action needed to address security challenges at chemical facilities.* GAO-04-482T. Washington, DC: Government Accountability Office.

———. 2005. *Wastewater facilities: Experts' view on how federal funds should be spent to improve security.* GAO-05-165. Washington, DC: Government Accountability Office.

Henry, K. 2002. New face of security. *Government Security* (2002): 30–37.

Mays, L. W. 2004. *Water supply systems security.* New York: McGraw-Hill.

Minter, J. G. 1996. Prevention of chemical accidents still a challenge. *Occupational Hazards,* September.

Schierow, L.-J. 2004. *Chemical plant security.* CRS Report for Congress RL31530. Washington, DC: Congressional Research Service.

Spellman, F. R. 1997. *A guide to compliance for PSM/RMP.* Lancaster, PA: Technomic Publishing.

U.S. Congress. Senate. 2005. *Wastewater treatment works security bill of 2005.* S 1995. 99th Cong.

U.S. Environmental Protection Agency (USEPA). 2002. *Vulnerability assessment fact sheet.* EPA 816-F-02-025. www.epa.gov/ogwdw/security/index.html (accessed May 2006).

6

Preparation: When Is Enough, Enough?

We must take the battle to the enemy, disrupt his plans, and confront the worst threats before they emerge.

—*George W. Bush*

Because of the seriousness of the threat of terrorism to the nation's chemical industry, the U.S. Environmental Protection Agency (USEPA) and other agencies have worked nonstop since 9/11 in gathering and providing as much advice and guidance as possible to aid chemical industry personnel in protecting chemical-processing facilities and associated critical support infrastructure. In this chapter, we provide an overview of one of USEPA's important tools in aiding the chemical industry to ward off and protect against the threat of terrorism. Though current material from the USEPA's (2003) *Response Protocol Toolbox* is provided in the following, it is important to note that the *Toolbox* is a living document—a work in progress—and is updated frequently.

THREATS AND INCIDENTS

Indication of the potential human and environmental losses and economic costs from an attack on a large chemical facility comes from major accidents that have occurred both abroad and in the United States. Those events indicate that the human and environmental losses could be significant (Congressional Budget Office 2004).

Chemical industry threats and incidents may be of particular concern due to the range of potential consequences:

- creating an adverse impact on public health within a population
- disrupting system operations and interrupting the supply of chemicals
- causing physical damage to system infrastructure

- reducing public confidence in the chemical supply system
- long-term denial of critical chemicals and the cost of replacement

Keep in mind that some of these consequences would only be realized in the event of a successful terrorist incident; however, the mere threat of terrorism can also have an adverse impact on industries that depend on a safe, steady supply of chemicals.

On the other hand, while it is important to consider the range of possibilities associated with a particular threat, assessments are typically based on the probability of a particular occurrence. Determining probability is somewhat subjective, and is often based on intelligence and previous incidents. As mentioned, there are historical accounts of accidental incidents that have caused tremendous death and destruction.

Threat Warning Signs

A threat warning is an occurrence or discovery that indicates a potential threat that triggers an evaluation of the threat. It is important to note that these warnings must be evaluated in the context of typical industry activity and previous experience in order to avoid false alarms. Following is a brief description of potential warnings:

- *Security breach.* Physical security breaches, such as unsecured doors, open hatches, and unlocked/forced gates, are probably the most common threat warnings. In most cases, the security breach is likely related to lax operations or typical criminal activity such as trespassing, vandalism, and theft. However, it may be prudent to assess any security breach with respect to the possibility of attack.
- *Witness account.* Awareness of an incident may be triggered by a witness account of tampering. Chemical industries should be aware that individuals observing suspicious behavior near chemical plants will likely call 911 and not the plant. In this case, the incident warning technically might come from law enforcement, as described below. Note: The witness may be a plant employee engaged in his or her normal duties.
- *Direct notification by perpetrator.* A threat may be made directly to the chemical plant, either verbally or in writing. Historical incidents would indicate that verbal threats made over the phone are more likely than written threats. While the notification may be a hoax, threatening a chemical plant is a crime and should be taken seriously.
- *Notification by law enforcement.* A chemical facility may receive notification about a threat directly from law enforcement, including local, county, state, or federal agencies. As discussed previously, such a threat could be a result of suspicious activity reported to law enforcement, either by a perpetrator, a witness, or the news media. Other information, gathered through intelligence or informants, could also lead law enforcement to conclude that there may be a threat to the chemical facility. While

law enforcement will have to lead in the criminal investigation, the chemical facility has primary responsibility for the safety of its chemicals and processes and public health. Thus, the plant's role will likely be to help law enforcement to appreciate the public health implications of a particular threat as well as the technical feasibility of carrying out a particular threat.

- *Notification by news media.* A threat to destroy a chemical facility might be delivered to the news media, or the media may discover a threat. A conscientious reporter would immediately report such a threat to the police, and either the reporter or the police would immediately contact the chemical facility. This level of professionalism would provide an opportunity for the plant to work with the media and law enforcement to assess the credibility of the threat before any broader notification is made.
- *Public health notification.* In this case, the first indication that chemical contamination or a chemical emergency (e.g., chemical burns) has occurred is the appearance of victims in local emergency rooms and health clinics. Chemical facilities may therefore be notified, particularly if the cause is unknown or linked to chemicals. An incident triggered by a public health notification is unique in that at least a segment of the population has been exposed to a harmful substance. If this agent is a chemical (including biotoxins), then the time between exposure and onset of symptoms may be on the order of hours, and thus there is the potential that the contaminant is still present.

RESPONSE TO THREATS

Note: This section is not designed to discuss what specific steps to take in responding to a contamination threat. Rather, the questions addressed in this section are "Why is it necessary to plan to respond to chemical industry threats at all?" and "When have I done enough?"

Federal, state, and local programs already exist that—with varying degrees of effectiveness—encourage or require the operators of chemical facilities to boost their efforts to promote safety and security and to share information that can help local governments plan for emergencies (Schierow 2006).

Proper planning is a delicate process because public health measures are rarely noticed or appreciated (like buried utility pipes, they are often hidden functions) except when they fail—then they are very visible. Consumers are particularly upset by unsafe (contaminated) environmental media (water, air, and/or soil) because they are often viewed as entitlements—and indeed, it is reasonable for consumers to expect a high-quality, safe environment. Public health failures during response to contamination threats often take the form of too much or too little action. The results of too little action, including no response at all, can have disastrous consequences potentially resulting in public injuries or fatalities. On the other hand, a disproportionate response to chemical contamination threats that have not been corroborated (i.e., determined to be credible) can also have serious repercussions when, for example, otherwise safe drinking water is unavailable because it is said to be chemically contaminated. Not only would the water be unavailable for human consumption, but it would also be unavailable for sanitation, firefighting, industry, and the many other uses of the public water supply. Although precise estimates are not available, information on accidents in the chemical industry, company assessment of what could happen in a severe release or explosion, and actual terrorist incidents involving chemicals suggest that the risk of attack is real and the losses could be significant. These adverse impacts must be considered when evaluating response options to a chemical contamination threat.

Considering the potential risks of an inappropriate response to a severe release or explosion threat, it is clear that a systematic approach is needed to evaluate chemical contamination threats. One overriding question is "When has a chemical industrial entity done enough?" This question may be particularly difficult to address when considering the wide range of agencies that may be involved in a threat situation. Other organizations, such as USEPA, CDC, DOT, law enforcement agencies, public health departments, and so on, will each have unique obligations or interests in responding to a severe release or explosion threat.

When Is Enough, Enough?

The guiding principle for responding to severe release or explosion threats is one of due diligence—or "What is a suitable and sensible response to a threat?" As discussed above, some response to chemical contamination threats is warranted due to the public health implications of an actual contamination incident. However, a chemical facility could spend a lot of time and money overresponding to every threat, which would be an ineffective use of resources. Furthermore, overresponse to threats carries its own adverse impacts.

Ultimately, the answer to the question of due diligence must be decided at the local level and will depend on a number of considerations. Among other factors, local authorities must decide what level of risk is reasonable in the context of a perceived threat. Careful planning is essential to developing an appropriate response to terrorist threats, and in fact, one primary objective of USEPA's *Response Protocol Toolbox* (RPTB) is to aid users in the development of their own site-specific plans that are consistent with the needs and responsibilities of the user. Beyond planning, the RPTB considers a careful evaluation of any terrorist threat, and an appropriate response based on the evaluation, to be the most important element of due diligence.

In the RPTB, the threat management process is considered in three successive stages: "possible," "credible," and "confirmed." Thus, as the threat escalates through these three states, the actions that might be considered due diligence expand accordingly. The following paragraphs describe, in general terms, actions that might be considered as due diligence at these various stages.

- *Stage 1: "Is the threat possible?"* If a chemical facility is faced with a terrorism threat, it should evaluate the available information to determine whether or not the threat is "possible" (i.e., could something have actually happened). If the threat is possible, immediate operational response actions might be implemented, and activities such as site characterization would be initiated to collect additional information to support the next stage of the threat evaluation.
- *Stage 2: "Is the threat credible?"* Once a threat is considered possible, additional information will be necessary to determine if the threat is "credible." The threshold at the credible stage is higher than that at the possible stage, and in general there must be information to corroborate the threat in order for it to be considered credible.
- *Stage 3: "Has the incident been confirmed?"* Confirmation implies that definitive evidence and information have been collected to establish the presence of a threat to the chemical facility. Obviously, at this stage the concept of due diligence takes on a whole new meaning since authorities are now faced with death and destruction and a potential public health crisis. Response actions at this point include all steps necessary to protect public health, property, and the environment.

PREPARATION

As environmental safety and health consultants for various utilities on the East Coast, we have performed numerous pre-OSHA audit inspections and audits of various plant PSM/RMP compliance programs. During these site visits, one factor seems to be universal. While conducting the plant walk-around to gauge the plant's overall profile and status with OSHA and USEPA compliance, we would almost always find that the plant manager or superintendent who accompanied us was shocked to find out what was actually going on in the facility. The manager or superintendent would scratch his or her head and ask various workers: "What the hell are you doing? Where did that new machine come from? When was it installed? Why is that door broken? Who told you to paint that door (or machine or other apparatus)? When did that hole get in the fence? Who left the back gate open? Where is the foreman?" . . . et cetera, et cetera, et cetera. Eventually, in an expression of utter consternation, the manager/superintendent would ask, "Who the hell is in charge around here?" And this is basically the question we find ourselves asking.

In one inspection performed at a plant right after 9/11, we drove up to the entrance gate and were impressed with the height and condition of the razor wire–topped fence and gates. We could not enter through the gates until we identified ourselves over a speaker system while a CCTV camera focused on my face. We were let in and given instructions to sign in at the main office. Not bad, just the way it should be—or so we thought at the time. After walking most of the plant site in the company of the plant manager, we approached the back fence area, which was close to a huge chlorine storage building. At the terminal end of the plant fence, we noticed a large gate that was propped open with ivy growing on it, through it, and around it—obviously, the gate had been in the open position for quite some time. We asked the plant manager why the gate was open. He stated that it was always open—that it led to a downhill path to a beach area below where plant personnel had constructed a picnic area, fronting the James River.

We walked the path to the bottom picnic area, looked around, and then looked back up the path toward the open gate and the prominent structure standing within, the chlorine storage building. While walking back up the path to the gate, we asked the plant manager if he was not concerned about the safety of the plant, because the gate was left open, and especially about the safety of the fifty tons of chlorine gas stored in the chlorine storage building.

"Nah, no way—we are safe here. I really don't see anyone swimming upriver just to get into the plant site, ha ha. Besides, we are surrounded by woods out here. There's nothing to attack, anyway."

Once inside the plant, we asked the plant manager if he was not worried about terrorists or some disgruntled former employee using a boat filled with explosives or some other weapon(s) gaining easy access to the plant and especially the chlorine building via the James River beach landing and picnic area?

"Nah, that will never happen. Who would be that stupid? There's nothing around here worth blowin' up!"

Later, when we checked the GIS system data and maps showing the plant and surrounding area, we noted that about one mile from the plant site was a large housing area, a brewery (with more than seven hundred employees), and a very large theme and historical park—visited by more than 1,500,000 people every summer.

Know Your Chemical Production System

All chemical industry managers and process equipment operating personnel must know their plant/facility. There is no excuse for not knowing every square inch of the plant site. In particular, plant workers should know about any and all construction activities under way on the site, the actual construction parameters of the plant, and especially the operation of all plant unit processes. In addition, plant management must know not only their operating staff but also their customers.

Construction and Operation

Each chemical industrial facility is unique with respect to age, operation, and complexity. Another aspect to be considered relates to the chemical processing system. This is important, particularly in evaluating the potential spread of a spilled contaminant. Propagation of chemical contaminants through a system is dependent on a number of factors, including mixing conditions at the point of contamination, hydraulic conditions within water systems at the time of the contaminant introduction, and reactions between the contaminant and other materials in the system or environment.

Information about construction materials used in the system may be contained within the plant records and can be useful in evaluating the fate and transport of a particular chemical contaminant through a system. For example, a particular contaminant may adsorb to the pipe material used in a utility's distribution system, and this type of information would be critical in evaluating remediation options following a chemical contamination incident.

Personnel

Employees of a chemical industrial plant are generally its most valuable asset in preparing for and responding to chemical contamination threats and incidents. They have knowledge of the system and of the hazards associated with the chemicals. The importance of knowledgeable and experienced personnel is highlighted by the complexity of most chemical processing systems. This complexity makes a successful contamination of a specific target contingent upon detailed knowledge of the system configuration and usage patterns. If perpetrators have somehow gained a sophisticated understanding of a chemical production process, the day-to-day experience of chemical system personnel will prove an invaluable tool to countering any attacks. For instance, personnel may continually look for unusual aspects of daily operation that might be interpreted as a potential threat warning, and may also be aware of specific characteristics of the system that make it vulnerable to chemical contamination or worse.

Customers

Knowledge of chemical customers is an important component of preventing and managing fires, explosions and/or contamination incidents. Prevention is based largely on understanding potential targets of chemicals. Of special concern may be hospitals, schools, government buildings, or other institutions where large numbers of people could be directly or indirectly affected by a hazardous chemical threat or incident. Steps taken to protect the customer's employees and property, such as enhancements to the physical security of the customer's property at these locations may deter the attack itself.

Chemical customers vary significantly with regard to their expectations of what constitutes acceptable service, so it is necessary to consider the manner in which chemicals are used in a particular system. For example, high chemical demand that is largely driven by industry has different implications compared to high usage rates in an urban center with a high population density. Some customers, such as factories and product manufacturers, may have certain chemical quality requirements. Sensitive sub-populations, including children and the elderly, can exhibit adverse health effects at doses more than an order of magnitude lower than those necessary to produce serious injury or death in a healthy adult. That being said, for the purposes of managing chemical fire, explosion, and contamination threats, it is important to keep in mind that the most important goal is protecting the health of the public as a whole. Planning, preparation, and allocation of resources should be directed toward protecting the public at large, beyond specific demographic groups or individual users.

Update ERP for Intentional Spill, Explosion, or Contamination

Emergency response plans (ERPs) are nothing new to chemical industries, since many have developed ERPs to deal with natural disasters, accidents, violence in the workplace, civil unrest, and so on. Because chemical industries are a vital part and ingredient of our way of life, it has been prudent for chemical industries to develop ERPs in order to help ensure the continuous flow of water to the community. However, many chemical industry ERPs developed prior to 9/11 do not explicitly deal with terrorist threats, such as intentional fire, explosion, or contamination. Recently, the U.S. Congress and federal regulators have required chemical industries to prepare or revise, as necessary, an ERP to reflect the findings of their vulnerability assessment and to address terrorist threats.

Establish Communication and Notification Strategy

Communication strategies must be planned and made available to all potential participants prior to an actual incident or threat. For the purposes of responding to a chemical industry threat, the communication structure could have several management levels within the industry, as well as external to the industry (governmental emergency response units, for example), that may be involved in management of a chemical industry threat. The hierarchy of potential participants includes the chemical industry, local government, regional government (i.e., county), state government, and federal government. Not all of these levels would necessarily be involved in every situation; however, the mechanism and process through which they interact must be decided in advance of an incident to achieve optimal public health and environmental protection. For any type of chemical incident, CHEMTREC (CHEMical TRansportation Emergency Center), which is dedicated to assisting emergency responders deal with incidents involving a hazardous material, is available twenty-four hours a day. Due to the number and variety of possible participants, planning for effective communication is critical.

Perform Training and Desk/Field Exercises

In addition to a lack of planning, another reason that emergency response plans fail is lack of training and practice. Training provides the necessary means for everyone involved to acquire the skills to fulfill his or her role during an emergency. It may also provide important buy-in to the response process from both management and staff, which is essential to the success of any response plan. Desk exercises (also known as "tabletops," "sand lot," or "dry runs") along with field exercises allow participants to practice their skills. Also, these exercises will provide a test of the plan itself, revealing strengths and weakness that may be used to improve the overall plan. Improvements can include measures not only for intentional explosion of chemicals or contamination of an environment with chemicals, but also for other emergencies faced by the chemical facility and the community at large.

Enhance Physical Security

Denying physical access to key sites within the water system may act as a deterrent to a perpetrator. Criminals often seek the easiest route of attack, just like a burglar prefers a house with an open window. Aside from deterring actual attacks, enhancing physical security has other benefits. For example, installation of fences and locks may reduce the rate of false alarms. Without surveillance equipment or locks, it may not be possible to determine whether a suspicious individual has actually entered a vulnerable area. The presence of a lock and a determination as to whether it has been cut or broken provides sound, although not definitive, evidence that an intrusion has occurred. Likewise, security cameras can be used to review security breaches and determine if the incident was simply due to trespassing or is a potential contamination threat. The costs of enhancing physical security may be justified by comparison to the cost of responding to just one credible chemical fire, explosion, or contamination threat involving site characterization and lab analysis for potential contaminants. Chapter 9 provides a more in-depth discussion of physical security devices.

Establish Baseline Monitoring Program

Background concentrations of suspected or tentatively identified hazardous chemical contaminants may be extremely important in determining if an incident has occurred. In some cases, and for some hazardous chemicals, background levels may be at detectable concentrations. Baseline occurrence information is derived from monitoring data and is used to characterize typical levels of a particular chemical contaminant.

SITE CHARACTERIZATION AND SAMPLING

Site characterization is defined as the process of collecting information from an investigation site in order to support the evaluation of a hazardous chemical–based incident that could lead to a fire, explosion, or contamination threat. Site characterization promotes and recommends the scientific method of inquiry, using the conceptual site modeling process to characterize the environmental condition of a site using a multiphased approach. Site characterization activities include the site evaluation, field safety screening, rapid field testing of the air and water, and sample collection. The investigation site is the focus of site characterization activities, and if a suspected contamination site has been identified, it will likely be designated as the primary investigation site. Additional or secondary investigation sites may be identified due to the potential spread of a suspected contaminant. The results of site characterization are of critical importance to the threat evaluation process.

There are two broad phases of site characterization: planning and implementation. The incident commander is responsible for planning, while the site characterization team is responsible for implementing the site characterization plan. This section is intended as a resource for those involved in either the planning or implementation phases of site characterization. While the target audience is primarily chemical industry managers and staff, other organizations (e.g., police, fire departments, FBI, and USEPA criminal investigators) may be involved in site characterization activities.

Site Characterization Process

USEPA's (2003) *Response Protocol Toolbox* for water contamination events is the model we use (because of its applicability) to describe the following five-stage chemical site characterization process:

1. *Customizing the site characterization plan.* A site characterization plan is developed for a specific threat (possibly from a generic site characterization plan) and guides the team during site characterization activities.
2. *Approaching the site.* Before entering the site, an initial assessment of site conditions and potential hazards is conducted at the site perimeter.
3. *Characterizing the site.* The customized site characterization plan is implemented by conducting a detailed site investigation and rapid testing of the water, air, and soil.
4. *Collecting samples.* Water/air/soil samples are collected in the event that lab analysis is required.
5. *Exiting the site.* Following completion of site characterization, the site is secured and personnel exit the site and undergo any necessary decontamination.

While site characterization can be considered and implemented as a discrete process, it is important to regard it as an element of the threat evaluation process. In particular, site characterization is an activity initiated in response to a possible accidental or intentional fire, explosion, or contamination threat in order to gather information to help determine whether or not the threat is credible. Initially, information from the threat evaluation supports the development of the customized site characterization plan. As this plan is implemented, the observations and results from site characterization feed into the threat evaluation. In turn, the revised threat evaluation may indicate that the threat is credible, not credible, or that the site characterization plan needs to be received in the field to collect more information in order to make this determination. Because threat evaluation and site characterization are interdependent, the incident commander must be in constant communication with the site characterization team while they are performing their tasks.

The first step is to develop a customized site characterization plan, which is based on the specific circumstances of the threat warning. This customized plan may be adapted from a generic site characterization plan, which is developed as part of a utility's preparation for responding to hazardous chemical threat. The site characterization team will use the customized plan as the basis for their activities at the investigation site. After an initial evaluation of available information, it is necessary to identify an investigation site where site characterization activities will be conducted. During the development of the customized plan, it is important to conduct an initial assessment of site hazards, which is critical to the safety of the site characterization team and may impact the makeup of the team. If there are obvious signs of hazards at the site, then teams trained in hazardous materials safety and handling techniques, such as HazMat, may need to conduct an initial hazard assessment at the site; the HazMat team may either clear the site for entry by utility personnel, or decide to perform all site characterization activities themselves. Obvious signs of hazards would provide a basis for determining that a threat is credible. Furthermore, the site might be considered a crime scene if there are obvious signs of hazards, and law enforcement may take over the site investigation.

Upon arrival at the site perimeter, the team first conducts field safety screening and observes site conditions. The purpose of field safety screening activities is to identify potential environmental hazards that might pose a risk to the site characterization team. The specific field safety screening performed should be identified in the site characterization plan, and might include screens for radioactivity and volatile organic chemicals (VOCs). If the team detects signs of hazard, they should stop their investigation and immediately contact the incident commander to report their findings.

If no immediate hazards are identified during the approach to the site, the incident commander will likely approve the team to enter the site and perform the site characterization. During this stage, the team will continue field safety screening at the site and conduct a detailed site investigation.

Rapid field testing has three objectives: (1) to provide additional information to support the threat evaluation process, (2) to provide tentative identification of contaminants that would need to be confirmed later by lab testing, and (3) to determine if hazards tentatively identified in the chemical release require special handling precautions. The specific rapid field testing performed should be identified in the site characterization plan. Specific field testing performed should be based on the circumstances of the specific threat and should consider the training, experience, and resources of the site characterization team. Negative field test results are not a reason to forgo chemical sampling, since field testing is limited in scope and can result in false negatives.

Following rapid field testing, samples of the potentially contaminated air/water/soil will be collected for potential lab analysis. The decision to send samples to a lab for analysis should be based on the outcome of the threat evaluation. If the threat is determined to be credible, then samples should be immediately delivered to the lab for analysis. The analytical approach for samples collected from the site should be developed with input from the supporting lab(s), based on information from the site characterization and threat evaluation.

At this point, response actions may be implemented to protect public health. However, if the threat is determined to be "not credible," then samples may be collected, preserved and stored in the event that it becomes necessary to analyze them later.

Upon completion of site characterization activities, the team should prepare to exit the site. At this stage, the team should make sure that they have documented their findings, collected all equipment and samples, and resecured the site (e.g., locked doors, hatches, and gates). If the site is considered to be a potentially hazardous site or crime scene, there may be additional steps involved in exiting the site.

Roles and Responsibilities

The incident commander and the site characterization team leader are key personnel in site characterization. The incident commander has overall responsibility for managing the response to the threat and is responsible for planning and directing site characterization activities. The incident commander may also approve the site characterization team to proceed with their activities at key decision points in the process (e.g., whether or not to enter the site following the approach).

The site characterization team leader is responsible for implementing the site characterization plan in the field and supervising site characterization personnel. The site characterization team leader must coordinate and communicate with the incident commander during site characterization.

Depending on the nature of the contamination threat, other agencies and organizations may be involved or otherwise assume some responsibility during planning and implementation of site characterization activities. However, the incident commander has the ultimate responsibility for determining the scope of the site characterization activities and the team makeup.

Planning for Site Characterization

Providing training of staff involved in site characterization and sampling activities is critical. Responding to the site of a potential contamination incident is very different from routine inspection and sampling activities performed by utility staff. The equipment and safety procedures used at the site of a potential contamination incident may differ significantly from those used during more typical field activities. Providing staff training in the procedures presented in this section will help to ensure that the procedures are properly and safely implemented during emergency situations.

Safety and Personnel Protection

Proper safety practices are essential for minimizing risk to the site characterization team and must be established prior to an incident in order to be effective. Field personnel involved in site characterization activities should have appropriate safety training to conform to appropriate regulations, such as OSHA 1910.120, which deals with hazardous chemical substances. If planners and field personnel do not conclude that these regulations are applicable to them, they may still wish to adopt some of the safety principles in these regulations. The following guidance is provided to help users develop their own safety policies and practices. These safety policies should be consistent with the equipment and capabilities of the site characterization team and any applicable regulations.

The appropriate level of personal protection necessary to safely perform the site characterization activities will depend on the assessment of site hazards that might pose a risk to the site characterization team. The hazard assessment may be further refined during the approach to the site, based on the results of the field safety screening and initial observations of site conditions. Two general scenarios are considered, one in which there are no obvious signs of immediate hazards, and one in which there are indicators of site hazards.

Sample Collection Kits and Field Test Kits

Two types of kits are discussed in this section, sample collection kits and field test kits. Sample collection kits will generally contain all sample containers, materials, supplies, and forms necessary to perform sample collection activities. Field test kits contain the equipment and supplies necessary to perform field safety screening and rapid field testing of the air, water, and/or soil. Sample collection kits will generally be less expensive to construct than field test kits. Sample collection kits can be pre-positioned throughout a system, while the more expensive field kits may be assigned to specific site characterization teams or personnel.

The design and construction of sample collection and field test kits is a planning activity, since these kits must be ready to go at a moment's notice in response to a "possible" contamination threat. In addition to improving the efficacy of the site characterization and sampling activities, advance preparation of sample collection and field test kits offers several advantages:

- Sample collection and field test kits can be standardized throughout an area to facilitate sharing of kits in the event of an emergency that requires extensive sampling.
- Collection of a complete sample set is more likely to be achieved through the use of predesigned kits.
- Sample collection kits can be pre-positioned at key locations to expedite the sampling process.
- Personnel responsible for site characterization can become familiar with the content of the kits and trained in the use of any specialized equipment.

Generic Site Characterization Plan

A site characterization plan is developed to provide direction and communication between the incident commander and the site characterization team, which will facilitate the safe and efficient implementation of site characterization activities. The plan should be developed expeditiously, since the site characterization results are an important input to the threat evaluation process. The rapid development of a site characterization plan can be facilitated by the development of a generic site characterization plan that is easily customized to a specific situation. While the circumstances of a particular threat warning will dictate the specifics of a customized site characterization plan, many activities and procedures will remain the same for most situations, and these common aspects can be documented in the generic site characterization plan. Potential elements of a generic plan include pre-entry criteria, communications, team organization and responsibilities, safety, field testing, sampling, and exiting the site.

Pre-entry criteria define the conditions and circumstances under which site characterization activities will be initiated and the manner in which these activities will proceed. At each stage of the process (i.e., approach to the site, on-site characterization activities, sample collection, and exiting the site), specific criteria may be defined for proceeding to the next stage. The pre-entry criteria may also specify the general makeup of the site characterization team under various circumstances. For example, under low-hazard conditions chemical facility teams may perform site characterization, while specially trained responders might be called upon to assist in the case of potentially hazardous conditions at the site. The criteria developed for a particular chemical facility should be consistent with the role that the facility has assumed in performing site characterization activities.

The generic plan should define communication processes to ensure rapid transmittal of findings and a procedure for obtaining approval to proceed to the next stage of site characterization. It is advisable for the site characterization team to remain in constant communication with the incident commander for the entire time that they are on-site. The plan should provide an approval process for the team to advance through the approach and on-site evaluation stages of the characterization, to ensure that the team is not advancing into a hazardous situation. Communication devices (e.g., cell phone, two-way radio, or panic button) can be used to alert incident command of problems/observations encountered in the field. The communication section of the generic plan should also discuss coordination with other agencies (e.g., law enforcement, fire department) and contingencies for contacting HazMat responders.

Field testing and sampling may be handled in the generic plan by presenting a menu that covers all potential options available to the utility, based on both internal and external capabilities. In developing a customized plan, the incident commander can simply check off the field tests and sampling requirements that are appropriate for the specific situation. The site characterization plan may also need to be revised in the field based on the observations of the team.

Quality Assurance for Field Testing and Sampling

Because of the diversity of potential field testing and sampling activities during the characterization, there may be no specific quality assurance (QA) activities that apply to all sampling procedures. However, the following general QA principles would apply in most cases and are consistent with the QA guidelines published by USEPA's Environmental Response Team:

- All data should be documented on field data sheets or within site logbooks.
- All instrumentation should be operated in accordance with operating instructions as supplied by the manufacturer, unless otherwise specified in the work plan. Equipment checkout and calibration activities should be done and documented prior to site characterization.
- Any relevant QA principles and plans specific to the particular facility or responding organization should be observed.

Maintaining Crime Scene Integrity

A suspected chemical contamination site that is the focus of site characterization activities could potentially become the scene of a criminal investigation. If law enforcement takes responsibility for incident command because they believe a crime has been committed, they will control the site and dictate how any additional activities, such as site characterization, are performed. In cases in which the chemical facility is still responsible for incident command, it may be prudent to take precautions to maintain the integrity of the potential crime scene during site characterization activities. The following guidelines for maintaining crime scene integrity are provided, although this should not necessarily be considered an exhaustive list:

- If there is substantial physical evidence of contamination at a site, the threat will likely be deemed credible from a utility and a law enforcement perspective. In this case, law enforcement may take control of the site and limit the activities performed by other organizations at the site.
- Substantial physical evidence of contamination might include discarded PPE, equipment (such as pumps and hoses), or containers with residual material. Special care should be taken to avoid moving or disturbing any potential physical evidence.
- Evidence should not be handled except at the direction of the appropriate law enforcement agency. Specially trained teams from the law enforcement community are best suited (and may be jurisdictionally required) for the collection of physical evidence from a contaminated crime scene.
- The collection of physical evidence is not generally considered time sensitive; however, site characterization and sampling activities *are* time sensitive due to the public health implications of contaminated environmental media: air, water, and/or soil. Thus, collection of environmental media samples may precede collection of physical evidence, and care must be taken not to disturb the crime scene while performing these activities. If samples can be collected outside of the boundaries of the suspected crime scene, this may avoid concerns about the integrity of the crime scene.
- Water, air, and/or soil samples collected for the purpose of confirming/dismissing a contamination threat and identifying a contaminant could potentially be considered evidence and should be handled accordingly.
- Since the analytical results may be considered evidence as well, it is important to use a qualified lab for analytical support. If law enforcement has taken control of the situation prior to sample collection, they may require the collection of an additional sample set to be analyzed by their designated lab.
- Photographs and videos can be taken during the site characterization for use in the criminal investigation. Law enforcement should be consulted for proper handling during and after taking photographs/videos to ensure integrity of the evidence.

Maintaining crime scene integrity during site characterization is largely an awareness issue. If the site characterization team integrates the guidelines outline above into their on-site activities, they will go a long way toward maintaining the integrity of the crime scene.

Customizing the Site Characterization Plan

The first stage of the site characterization process is the customization of the generic plan developed as part of planning and preparation for responding to contamination threats. In general, the incident commander will develop the customized plan in conjunction with the site characterization team leader. The steps involved in the development of the plan include (1) perform an initial evaluation of information about the threat, (2) identify one or more investigation sites, (3) assess potential site hazards, (4) develop a sampling approach, and (5) assemble a site characterization team.

Site Characterization Report

In order to provide useful information to support the threat evaluation process and the development of an analytical approach, the findings of the site characterization should be summarized in a report. This report is not intended to be a formal document, but simply a concise summary of information from the site activities that can be quickly assembled within an hour or two. The recommended content of the report includes

- general information about the site
- information about potential site hazards
- summary of observations from the site evaluation
- field safety screening results, including any appropriate caveats on the results
- rapid field water, air, and/or soil testing results, including any appropriate caveats on the results
- inventory of samples collected, and the sites from which they were collected
- any other pertinent information developed during the site characterization

Sample Packaging and Transport

In order to perform analysis of samples beyond rapid field testing, it will be necessary to properly package the samples for transport to the appropriate labs as quickly as possible. Prompt and proper packaging and transport of samples will

- protect the integrity of samples from changes in composition or concentration caused by bacterial growth or degradation that might occur at increased temperatures
- reduce the chance of leaking or breaking of sample containers that would result in loss of sample volume, loss of sample integrity, and potential exposure of personnel to hazardous substances
- help ensure compliance with shipping regulations

REFERENCES AND RECOMMENDED READING

American Water Works Association Research Foundation (AwwaRF). 2002. *Online monitoring for drinking water utilities.* Denver, CO: American Water Works Association.

Congressional Budget Office. 2004. *Homeland security and the private sector.* Washington, DC: Congressional Budget Office.

Government Accountability Office (GAO). 2004. *Homeland security—Chemical security.* GAO-04-482T. Washington, DC: Government Accountability Office.

Schierow, L.-J. 2006. *Chemical plant security.* CRS Report #RL31530. Washington, DC: Congressional Research Service.

U.S. Environmental Protection Agency (USEPA). 2001. *Protecting the nation's water supplies from terrorist attack: Frequently asked questions.* Washington, DC: U.S. Environmental Protection Agency.

———. 2003. *Response protocol toolbox: Planning for and responding to drinking water contamination threats and incidents.* Washington, DC: U.S. Environmental Protection Agency.

World Health Organization (WHO). 2003. *Public health response to biological and chemical weapons: WHO guidance.* 2nd ed. Geneva: World Health Organization.

7

SCADA and the Chemical Industry

On April 23, 2000, police in Queensland, Australia, stopped a car on the road and found a stolen computer and radio inside. Using commercially available technology, a disgruntled former employee had turned his vehicle into a pirate command center of sewage treatment along Australia's Sunshine Coast. The former employee's arrest solved a mystery that had troubled the Maroochy Shire wastewater system for two months. Somehow the system was leaking hundreds of thousands of gallons of putrid sewage into parks, rivers and the manicured grounds of a Hyatt Regency hotel—marine life died, the creek water turned black and the stench was unbearable for residents. Until the former employee's capture—during his 46th successful intrusion—the utility's managers did not know why.

Specialists study this case of cyber-terrorism because it is the only one known in which someone used a digital control system deliberately to cause harm. The former employee's intrusion shows how easy it is to break in—and how restrained he was with his power.

To sabotage the system, the former employee set the software on his laptop to identify itself as a pumping station, and then suppressed all alarms. The former employee was the "central control station" during his intrusions, with unlimited command of 300 SCADA nodes governing sewage and drinking water alike.

The bottom line: as serious as the former employee's intrusions were they pale in comparison with what he could have done to the fresh water system—he could have done anything he liked. (Gellman 2002)

IN THE WORDS OF MASTER SUN TZU
FROM *THE ART OF WAR*:

Those who are first on the battlefield and await the opponents are at ease; those who are last, and head into battle are worn out.

Table 7.1. Threats to critical infrastructure observed by the FBI

Threat	*Description*
Criminal groups	There is an increased use of cyber intrusions by criminal groups, who attack systems for purposes of monetary gain.
Foreign intelligence services	Foreign intelligence services use cyber tools as part of their information gathering and espionage activities.
Hackers	Hackers sometimes crack into networks for the thrill of the challenge or for bragging rights in the hacker community. While remote cracking once required a fair amount of skill or computer knowledge, hackers can now download attack scripts and protocols from the Internet and launch them against victim sites. Thus, while attack tools have become more sophisticated, they have also become easier to use.
Hacktivists	Hacktivism refers to politically motivated attacks on publicly accessible Web pages or e-mail servers. These groups and individuals overload e-mail servers and hack into Web sites to send a political message.
Information warfare	Several nations are aggressively working to develop information warfare doctrines, programs, and capabilities. Such capabilities enable a single entity to have a significant and serious impact by disrupting the supply, communications, and economic infrastructures that support military power—impacts that, according to the Director of Central Intelligence, can affect the daily lives of Americans across the country.
Inside threat	The disgruntled organization insider is a principal source of computer crimes. Insiders may not need a great deal of knowledge about computer intrusions because their knowledge of a victim system often allows them to gain unrestricted access to cause damage to the system or to steal system data. The insider threat also includes outsourcing vendors.
Virus writers	Virus writers are posing an increasingly serious threat. Several destructive computer viruses and "worms" have harmed files and hard drives, including the Melissa Macro Virus, the Explore.Zip worm, the CIH (Chernobyl) Virus, Nimda, and Code Red.

Source: FBI (2000).

In 2000, the FBI identified and listed threats to critical infrastructure. These threats are listed and described in table 7.1.

THE CHEMICAL INDUSTRY AND CYBERSPACE

In the past few years, especially since 9/11, it has been somewhat routine for us to pick up a newspaper or magazine or view a television news program where a major topic of discussion is cyber security or the lack thereof. Many of the cyber intrusion incidents we read or hear about have added new terms or new uses for old terms to our vocabulary. For example, old terms such as *Trojan horse*, *worm*, and *virus* have taken on new connotations in regard to cyber security issues. Relatively new terms such as *scanner*, *Windows NT hacking tool*, *ICQ hacking tool*, *mail bomb*, *sniffer*, *logic bomb*, *nuker*, *dot*, *backdoor Trojan*, *key logger*, *hackers' Swiss knife*, *password cracker*, and *BIOS cracker* are now commonly read or heard. New terms have evolved along with various control mechanisms. For example, because many control systems are vulnerable to attacks of varying degrees, these attack attempts range from telephone line sweeps (*wardialing*) to wireless network sniffing (*wardriving*), and physical network port scanning to physical monitoring and intrusion. When wireless network sniffing is performed at (or near) the target point by a pedestrian it is called *warwalking*, meaning that instead of a person being in an automotive vehicle, the potential intruder may be sniffing the network for weaknesses or vulnerabilities on foot, posing as a person walking, but they may have a handheld PDA device or laptop computer (Warwalking 2003).

Not all relatively new and universally recognizable cyber terms have sinister connotations or meanings, of course. Consider, for example, the following digital terms: *backup, binary, bit, byte, CD-ROM, CPU, database, e-mail, HTML, icon, memory, cyberspace, modem, monitor, network, RAM, Wi-Fi* (wireless fidelity), *record, software, World Wide Web*—none of these terms normally generates thoughts of terrorism in most of us.

There is, however, one digital term, *SCADA*, that most people have not heard of. This is not the case, however, with those who work with the nation's critical infrastructure, including chemical industry functions. SCADA, or Supervisory Control and Data Acquisition System (also sometimes referred to as digital control systems or process control systems), plays an important role in computer-based control systems. From coordinating music and lights in the proper sequence with spray from water fountains to controlling systems used in the drilling and refining of oil and natural gas, control systems perform many functions. Many chemical facilities use computer-based systems to remotely control sensitive processes and system equipment previously controlled manually. These systems (commonly known as SCADA) allow a chemical facility (or operation) to monitor tank levels, to ensure that ingredients are mixed in the proper proportions, and, as mentioned, to collect data from sensors and control equipment located at remote sites. Common chemical system sensors measure elements such as fluid level, temperature, pressure, water purity or slurry composition, water/chemical clarity and/or purity, and chemical pipeline flow rates. Common chemical industry system equipment includes valves, pumps, and mixers for mixing chemicals in process operations. The critical infrastructure of many countries is increasingly dependent on SCADA systems.

WHAT IS SCADA?

Simply, SCADA is a computer-based control system that remotely controls processes previously controlled manually. The philosophy behind SCADA control systems can be summed up by the phrase, "If you can measure it, you can control it." SCADA allows an operator using a central computer to supervise (control and monitor) multiple networked computers at remote locations. Each remote computer can control mechanical processes (mixers, pumps, valves, etc.) and collect data from sensors at its remote location. Thus the phrase Supervisory Control and Data Acquisition, or SCADA.

The central computer is called the master terminal unit, or MTU. The MTU has two main functions: to periodically obtain data from RTUs/PLCs and to control remote devices through the operator station. The operator interfaces with the MTU using software called human machine interface (HMI). The remote computer is called the program logic controller (PLC) or remote terminal unit (RTU). The RTU activates a relay (or switch) that turns mechanical equipment on and off. The RTU also collects data from sensors. Sensors perform measurement, and actuators perform control.

In the initial stages utilities ran wires, also known as hardwire or land lines, from the central computer (MTU) to the remote computers (RTUs). Since remote locations can be located hundreds of miles from the central location, utilities began to use public phone lines and modems, leased telephone company lines, and radio and microwave communication. More recently, they have also begun to use satellite links, the Internet, and newly developed wireless technologies.

DID YOU KNOW?

Modern RTUs typically use a ladder-logic approach to programming due to its similarity to standard electrical circuits. An RTU that employs this ladder-logic programming is called a programmable logic controller (PLC).

Since the SCADA systems' sensors provided valuable information, many utilities and other industries established connections between their SCADA systems and their business system. This allowed utility/industrial management and other staff access to valuable statistics, such as chemical usage. When utilities/industries later connected their systems to the Internet, they were able to provide stakeholders/stockholders with usage statistics on the utility/industrial web pages. Figure 7.1 provides a basic illustration of a representative SCADA network. Note that firewall protection (see chapter 9) would normally be placed between the Internet and the business system and between the business system and the MTU.

SCADA Applications in Chemical Industrial Systems

As stated above, SCADA systems can be designed to measure a variety of equipment operating conditions and parameters, or volumes and flow rates, or chemical and chemical mixture quality parameters, and to respond to changes in those parameters

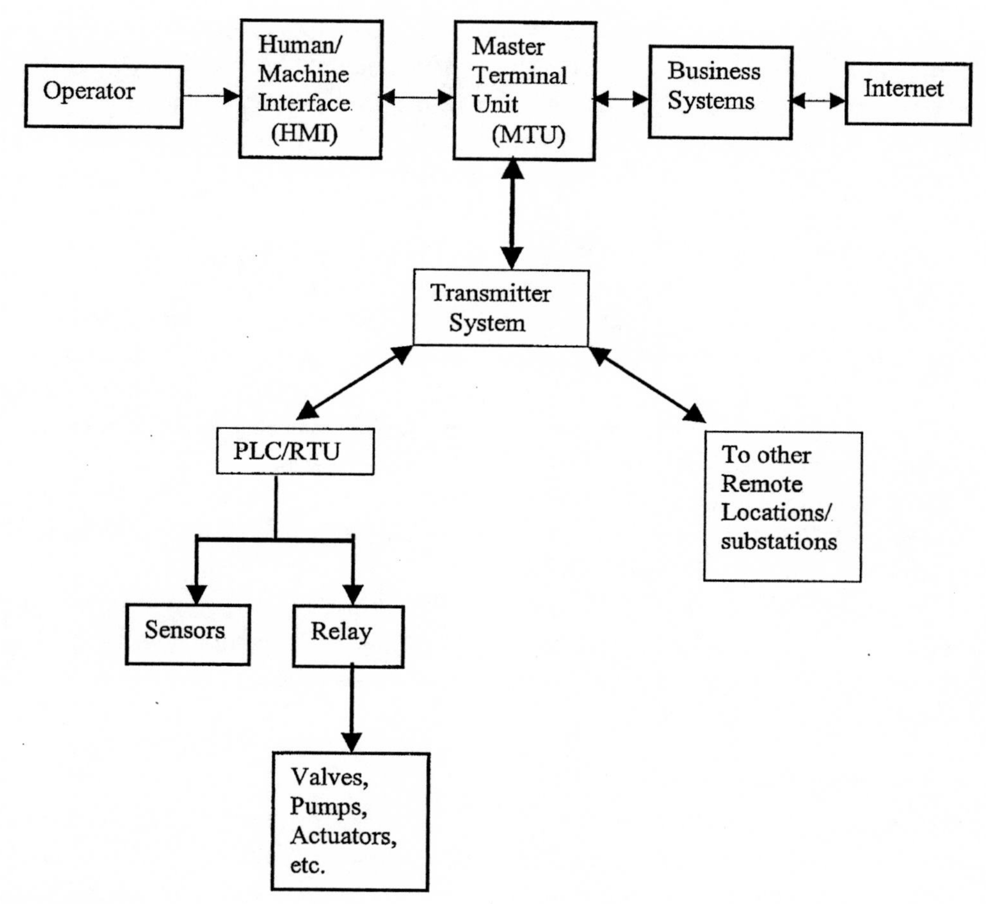

FIGURE 7.1
Representative SCADA network.

either by alerting operators or by modifying system operation through a feedback loop system, without having personnel physically visit each process or piece of equipment on a daily basis to check it and/or ensure that it is functioning properly. Automation and integration of large-scale diverse assets required SCADA systems to provide the utmost in flexibility, scalability, openness, and reliability. SCADA systems are used to automate certain chemical production functions that can be performed without needing to be initiated by an operator (e.g., injecting chlorine in response to periodic low chlorine levels in a water distribution system, or turning on a pump in response to low chemical levels in a storage tank). In addition to process equipment, SCADA systems can also integrate specific security alarms and equipment, such as cameras, motion sensors, lights, data from card-reading systems, and so on, thereby providing a clear picture of what is happening at areas throughout a facility. Finally, SCADA systems also provide constant, real-time data on processes, equipment, location access, and so forth, so the necessary response can be made quickly. This can be extremely useful during emergency conditions, such as when chemical distribution piping breaks or when potentially disruptive chemical reaction spikes appear in chemical processing operations.

Because these systems can monitor multiple processes, equipment, and infrastructure and then provide quick notification of, or response to, problems or upsets, SCADA systems typically provide the first line of detection for atypical or abnormal conditions. For example, a SCADA system that is connected to sensors that measure specific water quality parameters shows measurements outside of a specific range. A real-time customized operator interface screen could display and control critical systems monitoring parameters.

The system could transmit warning signals back to the operators, such as by initiating a call to a personal pager. This might allow the operators to initiate actions to prevent contamination and disruption of the water supply. Further automation of the system could ensure that the system initiated measures to rectify the problem. Preprogrammed control functions (e.g., shutting a valve, controlling flow, increasing chlorination, or adding other chemicals) can be triggered and operated based on a SCADA utility.

SCADA VULNERABILITIES

According to the U.S. Environmental Protection Agency (2005), SCADA networks were developed with little attention paid to security, making the security of these systems often weak. Studies have found that while technological advancements introduced vulnerabilities, many industrial facilities and utilities have spent little time securing their SCADA networks. As a result, many SCADA networks may be susceptible to attacks and misuse.

Remote monitoring and supervisory control of processes began to develop in the early 1960s, and adopted many technological advancements. The advent of minicomputers made it possible to automate a vast number of once manually operated switches. Advancements in radio technology reduced the communication costs associated with installing and maintaining buried cable in remote areas. SCADA systems continued to adopt new communication methods, including satellite and cellular. As the price of computers and communications dropped, it became economically feasible to distribute operations and to expand SCADA networks to include even smaller facilities.

Advances in information technology and the necessity of improved efficiency have resulted in increasingly automated and interlinked infrastructures, and have created new vulnerabilities due to equipment failure, human error, weather and other natural causes, and physical and cyber attacks. Some areas and examples of possible SCADA vulnerabilities include

- *Human.* People can be tricked or corrupted, and they may commit errors.
- *Communications.* Messages can be fabricated, intercepted, changed, deleted, or blocked.
- *Hardware.* Security features are not easily adapted to small, self-contained units with limited power supplies.
- *Physical.* Intruders can break into a facility to steal or damage SCADA equipment.
- *Natural.* Tornadoes, floods, earthquakes, and other natural disasters can damage equipment and connections.
- *Software.* Programs can be poorly written.

A study included a survey that found that many water utilities were doing little to secure their SCADA network vulnerabilities (Ezell 1998). For example, many respondents reported that they had remote access, which can allow an unauthorized person to access the system without being physically present. More than 60 percent of the respondents believed that their systems were not safe from unauthorized access and use. Twenty percent of the respondents even reported known attempts, successful unauthorized access, or use of their system. Yet twenty-two of forty-three respondents reported that they do not spend any time ensuring their network is safe, and eighteen of the forty-three respondents reported that they spend less than 10 percent ensuring network safety.

SCADA system computers and their connections are susceptible to different types of information system attacks and misuse, such as system penetration and unauthorized access to information. The Computer Security Institute and Federal Bureau of Investigation conduct an annual Computer Crime and Security Survey (Computer Security Institute 2004). The 2004 survey reported on ten types of attacks or misuse, and reported that viruses and denial of service had the greatest negative economic impact. The same study also found that 15 percent of the respondents reported abuse of wireless networks, which can be a SCADA component. On average, respondents from all sectors did not believe that their organization invested enough in security awareness. Utilities as a group reported a lower average computer security expenditure/investment per employee than many other sectors, such as transportation, telecommunications, and financial.

Sandia National Laboratories' *Common Vulnerabilities in Critical Infrastructure Control Systems* described some of the common problems it has identified in the following five categories (Stamp et al. 2003):

1. *System data.* Important data attributes for security include availability, authenticity, integrity, and confidentiality. Data should be categorized according to its sensitivity,

and ownership and responsibility must be assigned. However, SCADA data is often not classified at all, making it difficult to identify where security precautions are appropriate (for example, which communication links to secure, databases requiring protection, etc.).

2. *Security administration.* Vulnerabilities emerge because many systems lack a properly structured security policy (security administration is notoriously lax in the case of control systems), equipment and system implementation guides, configuration management, training, and enforcement and compliance auditing.
3. *Architecture.* Many common practices negatively affect SCADA security. For example, while it is convenient to use SCADA capabilities for other purposes such as fire and security systems, these practices create single points of failure. Also, the connection of SCADA networks to other automation systems and business networks introduces multiple entry points for potential adversaries.
4. *Network (including communication links).* Legacy systems' hardware and software have very limited security capabilities, and the vulnerabilities of contemporary systems (based on modern information technology) are publicized. Wireless and shared links are susceptible to eavesdropping and data manipulation.
5. *Platforms.* Many platform vulnerabilities exist, including default configurations retained, poor password practices, shared accounts, inadequate protection for hardware, and nonexistent security monitoring controls. In most cases, important security patches are not installed, often due to concern about negatively impacting system operation; in some cases technicians are contractually forbidden from updating systems by their vendor agreements.

The following incident helps to illustrate some of the risks associated with SCADA vulnerabilities.

> During the course of conducting a vulnerability assessment, a contractor stated that personnel from his company penetrated the information system of a utility within minutes. Contractor personnel drove to a remote substation and noticed a wireless network antenna. Without leaving their vehicle, they plugged in their wireless radios and connected to the network within five minutes. Within twenty minutes they had mapped the network, including SCADA equipment, and accessed the business network and data. This illustrates what a cyber security adviser from Sandia National Laboratories specializing in SCADA stated, that utilities are moving to wireless communication without understanding the added risks.

THE INCREASING RISK

According to the Government Accountability Office (2003), historically, security concerns about control systems (SCADA included) were related primarily to protecting against physical attack and misuse of refining and processing sites or distribution and holding facilities. However, more recently there has been a growing recognition that control systems are now vulnerable to cyber attacks from numerous sources, including hostile governments, terrorist groups, disgruntled employees, and other malicious intruders.

In addition to the control system vulnerabilities mentioned earlier, several factors have contributed to the escalation of risk to control systems, including (1) the adoption of standardized technologies with known vulnerabilities, (2) the connectivity of control systems to other networks, (3) constraints on the implementation of existing security technologies and practices, (4) insecure remote connections, and (5) the widespread availability of technical information about control systems.

Adoption of Technologies with Known Vulnerabilities

When a technology is not well known, not widely used, and not understood or publicized, it is difficult to penetrate it and thus disable it. Historically, proprietary hardware, software, and network protocols made it difficult to understand how control systems operated—and therefore how to hack into them. Today, however, to reduce costs and improve performance, organizations have been transitioning from proprietary systems to less expensive, standardized technologies such as Microsoft's Windows and Unix-like operating systems and the common networking protocols used by the Internet. These widely used standardized technologies have commonly known vulnerabilities, and sophisticated and effective exploitation tools are widely available and relatively easy to use. As a consequence, both the number of people with the knowledge to wage attacks and the number of systems subject to attack have increased. Also, common communication protocols and the emerging use of Extensible Markup Language (commonly referred to as XML) can make it easier for a hacker to interpret the content of communications among the components of a control system.

Control systems are often connected to other networks—enterprises often integrate their control systems with their enterprise networks. This increased connectivity has significant advantages, including providing decision makers with access to real-time information and allowing engineers to monitor and control the process control system from different points on the enterprise network. In addition, the enterprise networks are often connected to the networks of strategic partners and to the Internet. Further, control systems are increasingly using wide area networks and the Internet to transmit data to their remote or local stations and individual devices. This convergence of control networks with public and enterprise networks potentially exposes the control systems to additional security vulnerabilities. Unless appropriate security controls are deployed in the enterprise network and the control system network, breaches in enterprise security can affect the operation of a control system.

According to industry experts, the use of existing security technologies, as well as strong user authentication and patch management practices, are generally not implemented in control systems because control systems operate in real time, typically are not designed with cybersecurity in mind, and usually have limited processing capabilities.

Existing security technologies such as authorization, authentication, encryption, intrusion detection, and filtering of network traffic and communications require more bandwidth, processing power, and memory than control system components typically have. Because controller stations are generally designed to do specific tasks, they use low-cost, resource-constrained microprocessors. In fact, some devices in the electrical industry still use the Intel 8088 processor, introduced in 1978. Consequently, it is difficult to install existing security technologies without seriously degrading the performance of the control system.

Further, complex passwords and other strong password practices are not always used to prevent unauthorized access to control systems, in part because this could hinder a rapid response to safety procedures during an emergency. As a result, according to experts, weak passwords that are easy to guess, shared, and infrequently changed are reportedly common in control systems, including the use of default passwords or even no password at all.

In addition, although modern control systems are based on standard operating systems, they are typically customized to support control system applications. Consequently, vendor-provided software patches are generally either incompatible or cannot be implemented without compromising service, shutting down "always-on" systems or affecting interdependent operations.

Potential vulnerabilities in control systems are exacerbated by insecure connections. Organizations often leave access links—such as dial-up modems to equipment and control information—open for remote diagnostics, maintenance, and examination of system status. Such links may not be protected with authentication or encryption, which increases the risk that hackers could use these insecure connections to break into remotely controlled systems. Also, control systems often use wireless communications systems, which are especially vulnerable to attack, or leased lines that pass through commercial telecommunications facilities. Without encryption to protect data as it flows through these insecure connections or authentication mechanisms to limit access, there is limited protection for the integrity of the information being transmitted.

Public information about infrastructures and control systems is available to potential hackers and intruders. The availability of this infrastructure and vulnerability data was demonstrated by a university graduate student, whose dissertation reportedly mapped every business and industrial sector in the American economy to the fiber-optic network that connects them—using material that was available publicly on the Internet, none of which was classified. Many of the electric utility officials who were interviewed for the National Security Telecommunications Advisory Committee's Information Assurance Task Force's electric power risk assessment expressed concern over the amount of information about their infrastructure that is readily available to the public.

In the electric power industry, open sources of information—such as product data and educational videotapes from engineering associations—can be used to understand the basics of the electrical grid. Other publicly available information—including filings of the Federal Energy Regulatory Commission (FERC), industry publications, maps, and material available on the Internet—is sufficient to allow someone to identify the most heavily loaded transmission lines and the most critical substations in the power grid.

In addition, significant information on control systems is publicly available—including design and maintenance documents, technical standards for the interconnection of control systems and RTUs, and standards for communication among control devices—all of which could assist hackers in understanding the systems and how to attack them. Moreover, there are numerous former employees, vendors, support contractors, and other end users of the same equipment worldwide with inside knowledge of the operation of control systems.

Cyber Threats to Control Systems

There is a general consensus—and increasing concern—among government officials and experts on control systems about potential cyber threats to the control systems that govern our critical infrastructures. As components of control systems increasingly make critical decisions that were once made by humans, the potential effect of a cyber threat becomes more devastating. Such cyber threats could come from numerous sources, ranging from hostile governments and terrorist groups to disgruntled employees and other malicious intruders. Based on interviews and discussions with representatives throughout the electric power industry, the Information Assurance Task Force of the National Security Telecommunications Advisory Committee concluded that an organization with sufficient resources, such as a foreign intelligence service or a well-supported terrorist group, could conduct a structured attack on the electric power grid electronically, with a high degree of anonymity and without having to set foot in the target nation.

In July 2002, the National Infrastructure Protection Center (NIPC) reported that the potential for compound cyber and physical attacks, referred to as "swarming attacks," is an emerging threat to the U.S. critical infrastructure. As NIPC reports, the effects of a swarming attack include slowing or complicating the response to a physical attack.

For instance, a cyber attack that disabled the water supply or the electrical system in conjunction with a physical attack could deny emergency services the necessary resources to manage the consequences—such as controlling fires, coordinating actions, and generating light.

Control systems, such as SCADA, can be vulnerable to cyber attacks. Entities or individuals with malicious intent might take one or more of the following actions to successfully attack control systems:

- disrupt the operation of control systems by delaying or blocking the flow of information through control networks, thereby denying availability of the networks to control system operations
- make unauthorized changes to programmed instructions in PLCs, RTUs, or DCS controllers, change alarm thresholds, or issue unauthorized commands to control equipment, which could potentially result in damage to equipment (if tolerances are exceeded), premature shutdown of processes (such as prematurely shutting down transmission lines), or even disabling of control equipment
- send false information to control system operators either to disguise unauthorized changes or to initiate inappropriate actions by system operators
- modify the control system software, producing unpredictable results
- interfere with the operation of safety systems

In addition, in control systems that cover a wide geographic area, the remote sites are often unstaffed and may not be physically monitored. If such remote systems are physically breached, the attackers could establish a cyber connection to the control network.

SECURING CONTROL SYSTEMS

Several challenges must be addressed to effectively secure control systems against cyber threats. These challenges include (1) the limitations of current security technologies in securing control systems, (2) the perception that securing control systems may not be economically justifiable, and (3) the conflicting priorities within organizations regarding the security of control systems.

A significant challenge in effectively securing control systems is the lack of specialized security technologies for these systems. The computing resources in control systems that are needed to perform security functions tend to be quite limited, making it very difficult to use security technologies within control system networks without severely hindering performance.

Securing control systems may not be perceived as economically justifiable. Experts and industry representatives have indicated that organizations may be reluctant to spend more money to secure control systems. Hardening the security of control systems would require industries to expend more resources, including acquiring more personnel, providing training for personnel, and potentially prematurely replacing current systems that typically have a lifespan of about twenty years.

Finally, several experts and industry representatives indicate that the responsibility for securing control systems typically includes two separate groups: IT security personnel and control system engineers and operators. IT security personnel tend to focus on securing enterprise systems, while control system engineers and operators tend to be more concerned with the reliable performance of their control systems. Further, they indicate that, as a result, those two groups do not always fully understand each other's requirements nor collaborate to implement secure control systems.

STEPS TO IMPROVE SCADA SECURITY

The President's Critical Infrastructure Protection Board and the Department of Energy (DOE) have developed the steps outlined below to help organizations improve the security of their SCADA networks. DOE (2001) points out that these steps are not meant to be prescriptive or all-inclusive. However, they do address essential actions to be taken to improve the protection of SCADA networks. The steps are divided into two categories: specific actions to improve implementation, and actions to establish essential underlying management processes and policies.

Twenty-One Steps to Increase SCADA Security (DOE, 2001)

The following steps focus on specific actions to be taken to increase the security of SCADA networks:

1. Identify all connections to SCADA networks.

   Conduct a thorough risk analysis to assess the risk and necessity of each connection to the SCADA network. Develop a comprehensive understanding of all connections to the SCADA network, and how well these connections are protected. Identify and evaluate the following types of connections:
   - internal local area and wide area networks, including business networks
   - the Internet
   - wireless network devices, including satellite uplinks
   - modem or dial-up connections
   - connections to business partners, vendors, or regulatory agencies
2. Disconnect unnecessary connections to the SCADA network.

   To ensure the highest degree of security of SCADA systems, isolate the SCADA network from other network connections to as great a degree as possible. Any connection to another network introduces security risks, particularly if the connection creates a pathway from or to the Internet. Although direct connections with other networks may allow important information to be passed efficiently and conveniently, insecure connections are simply not worth the risk; isolation of the SCADA network must be a primary goal to provide needed protection. Strategies such as utilization of "demilitarized zones" (DMZs) and data warehousing can facilitate the secure transfer of data from the SCADA network to business networks. However, they must be designed and implemented properly to avoid introduction of additional risk through improper configuration.

3. Evaluate and strengthen the security of any remaining connections to the SCADA network.

   Conduct penetration testing or vulnerability analysis of any remaining connections to the SCADA network to evaluate the protection posture associated with these pathways. Use this information in conjunction with risk management processes to develop a robust protection strategy for any pathways to the SCADA network. Since the SCADA network is only as secure as its weakest connecting point, it is essential to implement firewalls, intrusion detection systems (IDSs), and other appropriate security measures at each point of entry. Configure firewall rules to prohibit access from and to the SCADA network, and be as specific as possible when permitting approved connections. For example, an Independent System Operator (ISO) should not be granted "blanket" network access simply because there is a need for a connection to certain components of the SCADA system. Strategically place IDSs at each entry point to alert security personnel of potential breaches of network security. Organization management must understand and accept responsibility for risks associated with any connection to the SCADA network.

4. Harden SCADA networks by removing or disabling unnecessary services.

   SCADA control servers built on commercial or open-source operating systems can be exposed to attack through default network services. To the greatest degree possible, remove or disable unused services and network demons to reduce the risk of direct attack. This is particularly important when SCADA networks are interconnected with other networks. Do not permit a service or feature on a SCADA network unless a thorough risk assessment of the consequences of allowing the service/feature shows that the benefits of the service/feature far outweigh the potential for vulnerability exploitation. Examples of services to remove from SCADA networks include automated meter reading/remote billing systems, e-mail services, and Internet access. An example of a feature to disable is remote maintenance. Numerous secure configuration guidelines for both commercial and open source operating systems are in the public domain, such as the National Security Agency's series of security guides. Additionally, work closely with SCADA vendors to identify secure configurations and coordinate any and all changes to operational systems to ensure that removing or disabling services does not cause downtime, interruption of service, or loss of support.

5. Do not rely on proprietary protocols to protect your system.

   Some SCADA systems use unique, proprietary protocols for communications between field devices and servers. Often the security of SCADA systems is based solely on the secrecy of these protocols. Unfortunately, obscure protocols provide very little "real" security. Do not rely on proprietary protocols or factor default configuration settings to protect your system. Additionally, demand that vendors disclose any backdoor or vendor interfaces to your SCADA systems, and expect them to provide systems that are capable of being secured.

6. Implement the security features provided by device and system vendors.

   Most older SCADA systems (most systems in use) have no security features whatsoever. SCADA system owners must insist that their system vendor implement security features in the form of product patches or upgrades. Some newer SCADA devices are shipped with basic security features, but these are usually disabled to ensure ease of installation.

   Analyze each SCADA device to determine whether security features are present. Additionally, factory default security settings (such as in computer network firewalls) are often set to provide maximum usability, but minimal security. Set all security features to provide the maximum level of security. Allow settings below maximum security only after a thorough risk assessment of the consequences of reducing the security level.

7. Establish strong controls over any medium that is used as a backdoor into the SCADA network.

   Where backdoor or vendor connections do exist in SCADA systems, strong authentication must be implemented to ensure secure communications. Modems, wireless, and wired networks used for communications and maintenance represent a significant vulnerability to the SCADA network and remote sites. Successful "war dialing" or "war driving" attacks could allow an attacker to bypass all other controls and have direct access to the SCADA network or resources. To minimize the risk of such attacks, disable inbound access and replace it with some type of callback system.

8. Implement internal and external intrusion detection systems and establish twenty-four-hour-a-day incident monitoring.

   To be able to effectively respond to cyber attacks, establish an intrusion detection strategy that includes alerting network administrators of malicious network activity originating from internal or external sources. Intrusion detection system monitoring is essential twenty-four hours a day; this capability can be easily set up through a pager. Additionally, incident response procedures must be in place to allow an effective response to any attack. To complement network monitoring, enable logging on all systems and audit system logs daily to detect suspicious activity as soon as possible.

9. Perform technical audits of SCADA devices and networks, and any other connected networks, to identify security concerns.

   Technical audits of SCADA devices and networks are critical to ongoing security effectiveness. Many commercial and open-source security tools are available that allow system administrators to conduct audits of their systems/networks to identify active services, patch level, and common vulnerabilities. The use of these tools will not solve systemic problems, but will eliminate the "paths of least resistance" that an attacker could exploit. Analyze identified vulnerabilities to determine their significance, and take corrective actions as appropriate. Track corrective actions and analyze this information to identify trends. Additionally, retest systems after corrective actions have been taken to ensure that vulnerabilities were actually eliminated. Scan nonproduction environments actively to identify and address potential problems.

10. Conduct physical security surveys and assess all remote sites connected to the SCADA network to evaluate their security.

    Any location that has a connection to the SCADA network is a target, especially unmanned or unguarded remote sites. Conduct a physical security survey and inventory access points at each facility that has a connection to the SCADA system. Identify and assess any source of information including remote telephone/computer network/fiber optic cables that could be tapped; radio and microwave links that are exploitable; computer terminals that could be accessed; and wireless local area network access points. Identify and eliminate single points of failure. The security of the site must be adequate to detect or prevent unauthorized access. Do not allow "live" network access points at remote, unguarded sites simply for convenience.

11. Establish SCADA "Red Teams" to identify and evaluate possible attack scenarios.

    Establish a "Red Team" to identify potential attack scenarios and evaluate potential system vulnerabilities. Use a variety of people who can provide insight into weaknesses of the overall network, SCADA systems, physical systems, and security controls. People who work on the system every day have great insight into the vulnerabilities of your SCADA network and should be consulted when identifying potential attack scenarios and possible consequences. Also, ensure that the risk from a malicious insider is fully evaluated, given that this represents one of the greatest threats to an organization. Feed information resulting from the "Red Team" evaluation into risk management processes to assess the information and establish appropriate protection strategies.

The following steps focus on management actions to establish an effective cyber security program:

12. Clearly define cyber security roles, responsibilities, and authorities for managers, system administrators, and users.

    Organization personnel need to understand the specific expectations associated with protecting information technology resources through the definition of clear and logical roles and responsibilities. In addition, key personnel need to be given sufficient authority to carry out their assigned responsibilities. Too often, good cyber security is left up to the initiative of the individual, which usually leads to inconsistent implementations and ineffective security. Establish a cyber security organizational structure that defines roles and responsibilities and clearly identifies how cyber security issues are escalated and who is notified in an emergency.
13. Document network architecture and identify systems that serve critical functions or contain sensitive information that requires additional levels of protection.

    Develop and document robust information security architecture as part of a process to establish an effective protection strategy. It is essential that organizations design their networks with security in mind and continue to have a strong understanding of their network architecture throughout its life cycle. Of particular importance, an in-depth understanding of the functions that the systems perform and the sensitivity of the stored information is required. Without this understanding, risk cannot be properly assessed and protection strategies may not be sufficient. Documenting the information security architecture and its components is critical to understanding the overall protection strategy, and identifying single points of failure.
14. Establish a rigorous, ongoing risk management process.

    A thorough understanding of the risks to network computing resources from denial-of-service attacks and the vulnerability of sensitive information to compromise is essential to an effective cyber security program. Risk assessments form the technical basis of this understanding and are critical to formulating effective strategies to mitigate vulnerabilities and preserve the integrity of computing resources. Initially, perform a baseline risk analysis based on a current threat assessment to use for developing a network protection strategy. Due to rapidly changing technology and the emergence of new threats on a daily basis, an ongoing risk assessment process is also needed so that routine changes can be made to the protection strategy to ensure it remains effective. Fundamental to risk management is identification of residual risk with a network protection strategy in place and acceptance of that risk by management.

15. Establish a network protection strategy based on the principle of defense-in-depth.

    A fundamental principle that must be part of any network protection strategy is defense-in-depth. Defense-in-depth must be considered early in the design phase of the development process, and must be an integral consideration in all technical decision making associated with the network. Utilize technical and administrative controls to mitigate threats from identified risks to as great a degree as possible at all levels of the network. Single points of failure must be avoided, and cyber security defense must be layered to limit and contain the impact of any security incidents. Additionally, each layer must be protected against other systems at the same layer. For example, to protect against the insider threat, restrict users to access only those resources necessary to perform their job functions.
16. Clearly identify cyber security requirements.

    Organizations and companies need structured security programs with mandated requirements to establish expectations and allow personnel to be held accountable. Formalized policies and procedures are typically used to establish and institutionalize a cyber security program. A formal program is essential for establishing a consistent, standards-based approach to cyber security throughout an organization and eliminates sole dependence on individual initiative. Polices and procedures also inform employees of their specific cyber security responsibilities and the consequences of failing to meet those responsibilities. They also provide guidance regarding actions to be taken during a cyber security incident and promote efficient and effective actions during a time of crisis. As part of identifying cyber security requirements, include user agreements and notification and warning banners. Establish requirements to minimize the threat from malicious insiders, including the need for conducting background checks and limiting network privileges to those absolutely necessary.
17. Establish effective configuration management processes.

    A fundamental management process needed to maintain a secure network is configuration management. Configuration management needs to cover both hardware configurations and software configurations. Changes to hardware or software can easily introduce vulnerabilities that undermine network security. Processes are required to evaluate and control any change to ensure that the network remains secure. Configuration management begins with well-tested and documented security baselines for your various systems.
18. Conduct routine self-assessments.

    Robust performance evaluation processes are needed to provide organizations with feedback on the effectiveness of cyber security policy and technical implementation. A sign of a mature organization is one that is able to self-identify issues, conduct root cause analyses, and implement effective corrective actions that address individual and systemic problems. Self-assessment processes that are normally part of an effective cyber security program include routine scanning for vulnerabilities, automated auditing of the network, and self-assessments of organizational and individual performance.

19. Establish system backups and disaster recovery plans.

    Establish a disaster recovery plan that allows for rapid recovery from any emergency (including a cyber attack). System backups are an essential part of any plan and allow rapid reconstruction of the network. Routinely exercise disaster recovery plans to ensure that they work and that personnel are familiar with them. Make appropriate changes to disaster recovery plans based on lessons learned from exercises.

20. Senior organizational leadership should establish expectations for cyber security performance and hold individuals accountable for their performance.

    Effective cyber security performance requires commitment and leadership from senior managers in the organization. It is essential that senior management establish an expectation for strong cyber security and communicate this to their subordinate managers throughout the organization. It is also essential that senior organizational leadership establish a structure for implementation of a cyber security program. This structure will promote consistent implementation and the ability to sustain a strong cyber security program. It is then important for individuals to be held accountable for their performance as it relates to cyber security. This includes managers, system administrators, technicians, and users/operators.

21. Establish policies and conduct training to minimize the likelihood that organizational personnel will inadvertently disclose sensitive information regarding SCADA system design, operations, or security controls.

    Release data related to the SCADA network only on a strict, need-to-know basis, and only to persons explicitly authorized to receive such information. "Social engineering," the gathering of information about a computer or computer network via questions to naïve users, is often the first step in a malicious attack on computer networks. The more information revealed about a computer or computer network, the more vulnerable the computer/network is. Never divulge data revealed to a SCADA network, including the names and contact information about the system operators/administrators, computer operating systems, and/or physical and logical locations of computers and network systems over telephones or to personnel unless they are explicitly authorized to receive such information. Any requests for information by unknown persons need to be sent to a central network security location for verification and fulfillment. People can be a weak link in an otherwise secure network. Conduct training and information awareness campaigns to ensure that personnel remain diligent in guarding sensitive network information, particularly their passwords.

REFERENCES AND RECOMMENDED READING

Brown, A. S. 2008. SCADA vs. the hackers. American Society of Mechanical Engineers. www .memagazine.org/backissues/dec02/features/scadavs/ (accessed May 10, 2008).

Computer Security Institute and Federal Bureau of Investigation. 2004. *Ninth annual computer crime and security survey.* San Francisco: Computer Security Institute.

Ezell, B. C. 1998. Risks of cyber attack to supervisory control and data acquisition for water supply. PhD diss., University of Virginia.

Federal Bureau of Investigation (FBI). 2000. *Threats to critical infrastructure.* Washington, DC: Federal Bureau of Investigation.

Gellman, B. 2002. Cyber-attacks by Al Qaeda feared: Terrorists at threshold of using Internet as tool of bloodshed, experts say. *Washington Post,* June 27, A01.

Government Accountability Office (GAO). 2003. *Critical infrastructure protection: Challenges in securing control systems.* Washington, DC: Government Accountability Office.

National Infrastructure Protection Center (NIPC). 2002. *National Infrastructure Protection Center report.* Washington, DC: National Infrastructure Protection Center.

Stamp, J., et al., 2003. *Common vulnerabilities in critical infrastructure control systems.* 2nd ed. Albuquerque, NM: Sandia National Laboratories.

U.S. Department of Energy (DOE). 2001. *21 steps to improve cyber security of SCADA networks.* Washington, DC: Department of Energy.

U.S. Environmental Protection Agency (USEPA). 2005. EPA needs to determine what barriers prevent water systems from securing known SCADA vulnerabilities. In *Final briefing report,* ed. J. Harris. Washington, DC: U.S. Environmental Protection Agency.

Warwalking. 2003. warwalking.tribe.net (accessed May 9, 2008).

Young, M. A. 2004. *SCADA systems security.* Bethesda, MD: SANS Institute.

8

Emergency Response

We're in uncharted territory.

—*Rudy Giuliani, September 11, 2001*

To secure ourselves against defeat lies in our own hands, but the opportunity of defeating the enemy is provided by the enemy himself.

—*Sun Tzu, The Art of War*

When New York Mayor Rudy Giuliani made the above statement to Police Commissioner Bernard Kerik at the World Trade Center site on September 11, 2001, to a point and to a degree, one of the first (and not to be forgotten) gross understatements of the twenty-first century had been uttered. Indeed, for citizens of the United States of America, the 9/11 events placed our level of consciousness, awareness, fear, and questions of what to do next in "uncharted territory." Actually, when you get right down to it, 9/11 generated more questions than anything else. Many are still asking the following questions today:

Why?

Why would anyone have the audacity to attack the United States?

What kind of cold-blooded killers would even think of conducting such an event?

Who were those Islamic radicals who perpetrated 9/11?

What were the terrorists' goals?

Why?

Why were we not ready for such an attack?

Why had we not foreseen such an event?

Why were our emergency responders so undermanned, ill-prepared, and ill-equipped to handle such a disaster?

What took the military fighter planes so long to respond?

What did our government really know (if anything) before the events occurred?

Could anyone have prevented it?

Bottom line questions: Why us? Hell, why anyone?

Why?

These and several other questions continue to resonate today; no doubt they will continue to haunt us for some time to come.

Maybe we ask post-9/11-related questions because of who we are, what we are, and what we are not. That is, because we are Americans we are free, uninhibited thinkers who think what we say and say what we think—isn't America great! Most Americans are soft-hearted and sympathetic to those in need—compassion is the very nature and soul of being American. Americans are not born terrorists; they are not born into a terrorist regime; they are not raised with fear in their hearts—they are not afraid every time they leave their homes and go about their daily business. Suicide bombers and other like terrorists are those that occupy some other faraway place, definitely not America, and they are definitely not American. Right?

Notwithstanding exceptions to the rule, such as Timothy McVeigh (a so-called red-blooded American, born and raised in America) and that other idiot (whether a national or foreigner) who mailed the anthrax, terrorism was foreign to us.

Today, from a safety/security point of view, based on the events of 9/11 and the anthrax events, we should no longer be asking why. Instead, we should not waste our time, money, and energy asking why or pointing a finger of blame at our government, military, 9/11 emergency responders, and/or the terrorists. We should stop asking why and shift our mindset to asking "what if." The point is that we need to stop feeling sorry for ourselves and accept the fact that there are folks out there who do not share our view of the American way of life. In chapter 6, in regard to security preparedness, we pointed to the need to ask what-if questions. It was pointed out that what-if analysis is a proactive approach used to prevent or mitigate certain disasters, extreme events—those human- or nature-generated. Obviously, asking and properly answering what-if questions has little effect on preventing the actions of Mother Nature, such as earthquakes, tornadoes, hurricanes (Katrina-type events), and others. On the other hand, it is true that what-if questions, when properly posed and answered (with results), can reduce the death toll and overall damage caused by these natural disasters. We are certainly aware that these natural events are possible, probable, and likely, and their effects can be horrendous—beyond tragic. The irony is apparent, however, especially when we ask how many of us are actually willing to move away from or out of earthquake zones, hurricane and tornado alleys, and floodplains to live somewhere else.

The fact is that we do not possess a crystal ball to foretell the future. What-if questions prepare us to react and respond to certain contingencies. And respond we must, because there are certain events we simply can't prevent. The best response to an event we can't prevent is summed up by the Boy Scout motto: Be prepared!

CHEMICAL INDUSTRY CONTINGENCY PLANNING

Emergency response planning—or contingency planning—for extreme events has long been standard practice for safety professionals in chemical industry systems operations. For many years, prudent practices have required consideration of the potential impact of severe natural events (forces of nature), including earthquakes, tornadoes, volcanoes, floods, hurricanes, and blizzards. These possibilities have been included in chemistry industry infrastructure emergency preparedness and disaster response planning. In addition, many chemical production facilities have considered the potential consequences of violence in the workplace. Today, as this text has pointed out, there is a new focus of concern: the potential effects of intentional acts by domestic (homegrown) or international (foreign) terrorists.

As a result, the security paradigm has not necessarily changed, but instead has been adjusted—reasonable, necessary, and sensible accommodations have been and continue to be made. Because we cannot foresee all future intentional acts of terrorism, we must be prepared to shift from the proactive to reactive mode on short notice—in some cases, on very short notice. Accordingly, we must be prepared to respond to, react to, and mitigate what we can't prevent.

The following criteria have not been established as anything other than guidelines and are offered not as definitive or official regulations but rather as informed advice (based on more than thirty years of experience) insofar as the subject matter is specific to both public and private sectors.

EMERGENCY RESPONSE PLANNING: STANDARD TEMPLATE

The goals of an emergency response plan (ERP) are to document and understand the steps needed to

- rapidly restore chemical production processing services after an emergency
- minimize chemical production process equipment damage
- minimize impact and loss to customers
- minimize negative impacts on public health and employee safety
- minimize adverse effects on the environment
- provide emergency public information concerning customer service
- provide hazardous chemical information for first responders and other outside agencies

Although we are concerned with the chemical industry in this text, the USEPA-developed *Large Water System Emergency Response Plan Outline: Guidance to Assist Community Water Systems in Complying with the Public Health Security and Bioterrorism Preparedness and Response Act of 2002* (dated July 2003), with minor adjustments, can be applied as a template for the chemical industry. This template provides guidance and recommendations to aid facilities in the preparation of emergency response plans under PL 107-188. The template is provided below.

Chemical Industry ERP Template

I. Introduction

Safe and reliable operation is vital to every industrial operation. Emergency response planning is an essential part of managing a chemical industrial process. The introduction should identify the requirement to have a documented emergency response plan (ERP), the goal(s) of the plan (e.g., be able to quickly identify an emergency and initiate timely and effective response action, be able to quickly respond, and be able to repair damages to minimize system downtime), and how access to the plan is limited. Plans should be numbered for control. Recipients should sign and date a statement that includes (1) their ERP number, (2) an agreement not to reproduce the ERP, and (3) that they have read the ERP.

ERPs do not necessarily need to be one document. They may consist of an overview document, individual emergency action procedures, checklists, additions to existing operations manuals, appendices, etc. There may be separate, more detailed plans for specific incidents. There may be plans that do not include particularly sensitive information and those that do. Existing applicable documents should be referenced in the ERP (e.g., chemical Risk Management Program, contamination response).

II. Emergency Planning Process

A. Planning partnerships

The planning process should include those parties who will need to help the chemical operation in an emergency situation (e.g., first responders, law enforcement, public health officials, nearby utilities, local emergency planning committees, testing labs, etc.). Partnerships should track from the chemical operation up through local, state, regional, and federal agencies, as applicable and appropriate, and could also document compliance with governmental requirements.

B. General emergency response policies, procedures, actions, documents

A short synopsis of the overall emergency management structure, how other industrial emergency response, contingency, and risk management plans fit into the ERP for chemical emergencies, and applicable policies, procedures, actions plans, and reference documents should be cited. Policies should include interconnect agreements with adjacent communities and just how the ERP may affect them.

C. Scenarios

Use your vulnerability assessment (VA) findings to identify specific emergency action steps required for response, recovery, and remediation for applicable incident types. In Section V of this plan, specific emergency actions procedures addressing each of the incident types should be addressed.

III. Emergency Response Plan—Policies

A. System-specific information

In an emergency, chemical industries need to have basic information for system personnel and external parties such as law enforcement, emergency responders, repair contractors/vendors, the media, and others. The information needs to be clearly formatted and readily accessible so system staff can find and distribute it quickly to those who may be involved in responding to the emergency. Basic information that may be presented in the emergency response plan are the system's ID number, system name, system address or location, directions to the system, population served, number of service connections, system owner, and information about the person in charge of managing the emergency. Distribution maps, detailed plant drawings, site plans, source/storage chemical locations, and operations manuals may be attached to this plan as appendices or referenced.

1. Chemical industry ID, owner, contact person
2. Population served and service connections
3. System components
    a) Pipes and constructed conveyances
    b) Physical barriers
    c) Isolation valves
    d) Chemical treatment, storage and distribution facilities
    e) Electronic, computer, or other automated systems that are utilized by the chemical industry
    f) Emergency power generators (on-site and portable)
    g) The use, storage, or handling of various chemicals
    h) The operation and maintenance of such system components

B. Chain-of-command chart developed in coordination with local emergency planning committee (internal and/or external emergency responders, or both)

1. Contact name
2. Organization and emergency response responsibility
3. Telephone number(s) (hardwire, cell phones, faxes, e-mail)
4. State twenty-four-hour emergency communications center telephone

C. Communications procedures: who, what, when

During most emergencies, it will be necessary to quickly notify a variety of parties both internal and external to the chemical plant. Using the Chain-of-Command Chart and all appropriate personnel from the lists below, indicate who activates the plan, the order in which notification occurs, and the members of the emergency response team. All contact information should be available for routine updating and readily available. The following lists are not intended to be all inclusive—they should be adapted to your specific needs.

1. Internal notification lists
   a) Operations dispatch
   b) Chemical plant manager
   c) Chemical processing manager
   d) Chemical storage/distribution manager
   e) Facility managers
   f) Chief chemical engineer
   g) Director of engineering
   h) Data (IT) manager
   i) Maintenance manager
   j) Other
2. Local notification
   a) Head of local government (mayor, city manager, chairman of board, etc.)
   b) Public safety officials—fire, local law enforcement (LLE), police, EMS, safety (If a malevolent act is suspected, LLE should be immediately notified and in turn will notify the FBI, if required. The FBI is the primary agency for investigating sabotage to any system or terrorist incidents.)
   c) Other government entities: health, schools, parks, finance, electric, etc.
3. External notification lists
   a) State Department of Environmental Quality (DEQ)
   b) USEPA
   c) State police
   d) State health department (lab)
   e) Critical customers (special considerations for hospitals, federal, state, and country government centers, etc.)
   f) Service aid
   g) Mutual aid

h) Chemical Information Sharing and Analysis Center (ISAC)/ CHEMTREC)

i) Commercial customers not previously notified

4. Public/media notification: when and how to communicate

Effective communication is a key element of emergency response, and a media or communications plan is essential to good communications. Be prepared by organizing basic facts about the crisis and your chemical system. Develop key messages to use with the media that are clear, brief, and accurate. Make sure your messages are carefully planned and have been coordinated with local and state officials. Considerations should be given to establishing protocols for both field and office staff to respectfully defer questions to the proper spokesperson.

Be prepared to list geographic boundaries of the affected area (e.g., west of Highway A, east of Highway B, north of Highway C, and south of Highway D, to ensure the public clearly understands the system boundaries).

D. Personnel safety

This should provide direction as to how operations staff, emergency responders, and the public should respond to a potential toxic release (e.g., chemical plumes released), including facility evacuation, personnel accountability, proper personal protective equipment (PPE) as dictated by the Risk Management Program and Process Safety Management Plan, and whether the nearby public should be "in-place sheltered" or evacuated.

E. Equipment

The ERP should identify equipment that can obviate or significantly lessen the impact of terrorist attacks or other intentional actions on the public health and protect the safety and supply of communities and individuals. The chemical facility should maintain an updated inventory of current equipment and repair parts for normal maintenance work.

Because of the potential for extensive or catastrophic damage that could result from a malevolent act, additional equipment sources should be identified for the acquisition and installation of equipment and repair parts in excess of normal usage. This should be based on the results of the specific scenarios and critical assets identified in the vulnerability assessment that could be destroyed. For example, numerous pumps, vats, and mixers, specifically designed for the chemical industry, could potentially be destroyed. A certain number of "long-lead" procurement equipment should be inventoried and the vendor information for such unique and critical equipment maintained. In addition, mutual aid agreements with other industries, and the equipment available under the agreement, should be addressed. Inventories of current equipment, repair parts, and associated vendors should be indicated under Item 29 "Equipment Needs/Maintenance of Equipment" of Section IV "Emergency Action Procedures."

G. Property protection

A determination should be made as to what chemical processing operation/facility should be immediately "locked down," which specific access control procedures implemented, what initial security perimeter established, and whether a possible secondary malevolent event may occur. The initial act may be a diversionary act.

H. Training, exercises, and drills

Emergency response training is essential. The purpose of the training program is to inform employees of what is expected of them during an emergency situation. The level of training on an ERP directly affects how well a chemical facility's employees can respond to an emergency. This may take the form of orientation scenarios, table-top workshops, functional exercises, etc.

I. Assessment

To evaluate the overall ERP's effectiveness and to ensure that procedures and practices developed under the ERP are adequate and are being implemented, the chemical industry staff should audit the program on a periodic basis.

IV. Emergency Action Procedures (EAPs)

These are detailed procedures used in the event of an operational emergency or malevolent act. EAPs may be applicable across many different emergencies and are typically common core elements of the overall municipality ERP (e.g., responsibilities, notifications lists, security procedures, etc.) and can be referenced.

A. Event classification/severity of emergency
B. Responsibilities of emergency director
C. Responsibilities of Incident Commander
D. Emergency Operations Center (EOC) activation
E. Division internal communications and reporting
F. External communications and notifications
G. Emergency telephone list (division internal contacts)
H. Emergency telephone list (off-site responders, agencies, state twenty-four-hour emergency phone numbers, and others to be notified)
I. Mutual aid agreements
J. Contact list of available emergency contractor services/equipment

K. Emergency equipment list (including inventory for each facility)
L. Security and access control during emergencies
M. Facility evacuation and lockdown and personnel accountability
N. Treatment and transport of injured personnel (including chemical exposure)
O. Chemical records—to compare against historical results for base line
P. List of available labs for emergency use
Q. Emergency chemical sampling and analysis
R. Water use restrictions during emergencies
S. Alternate temporary chemical supplies during emergencies
T. Isolation plans for chemical supply, treatment, storage, and distribution systems
U. Mitigation plans for neutralizing, flushing, and collecting spilled chemicals
V. Protection of vital records during emergencies
W. Record keeping and reporting (FEMA, DHS, DOT, OSHA, EPA, and other requirements) (It is important to maintain accurate financial records of expenses associated with the emergency event for possible federal reimbursement.)
X. Emergency program training, drills, and tabletop exercises
Y. Assessment of emergency management plan and procedures
Z. Crime scene preservation training and plans
AA. Communication plans:
    1. Police
    2. Fire
    3. Local government
    4. Media
    5. Etc.
BB. Administration and logistics, including EOC, when established
CC. Equipment needs/maintenance of equipment
DD. Recovery and restoration of operations
EE. Emergency event closeout and recovery

V. Incident-Specific Emergency Action Procedures (EAPs)

Incident-Specific EAPs are action procedures that identify specific steps in responding to an operational emergency or malevolent act.

A. General response to terrorist threats (other than bomb threat and incident-specific threats)
B. Incident-specific response to man-made or technological emergencies
    1. Contamination event (articulated threat with unspecified materials)

      2. Contamination threat at a major event
      3. Notification from health officials of potential contamination
      4. Intrusion through Supervisory Control and Data Acquisition (SCADA)
   C. Significant structural damage resulting from intentional act
   D. Customer complaints
   E. Severe weather response (snow, ice, temperature, lightning)
   F. Flood response
   G. Hurricane and/or tornado response
   H. Fire response
   I. Explosion response
   J. Major vehicle accident response
   K. Electrical power outage response
   L. Water supply interruption response
   M. Transportation accident response—barge, plane, train, semi-trailer/tanker
   N. Contaminated/tampered with water treatment chemicals
   O. Earthquake response
   P. Disgruntled employees response (i.e., workplace violence)
   Q. Vandals response
   R. Bomb threat response
   S. Civil disturbance/riot/strike
   T. Armed intruder response
   U. Suspicious mail handling and reporting
   V. Hazardous chemical spill/release response (including Material Safety Data Sheets)
   W. Cyber-security/Supervisory Control and Data Acquisition (SCADA) system attack response (other than incident-specific, e.g., hacker)

VI. Next Steps
   A. Plan review and approval
   B. Practice and plan to update (as necessary, once every year recommended)
      1. Training requirements
      2. Who is responsible for conducting training, exercises, and emergency drills
      3. Update and assessment requirements
      4. Incident-specific requirements

VII. Annexes
   A. Facility and location information
      1. Facility maps
      2. Facility drawings

   3. Facility descriptions/layout
   4. Etc.

VIII. References and Links
   A. Department of Homeland Security. www.dhs/gov/dhspublic
   B. Environmental Protection Agency. www.epa.gov
   C. American Water Works Association (AWWA). www.awwa.org
   D. Centers for Disease Control and Prevention. www.bt.cdc.gov
   E. Federal Emergency Management Agency. www.fema.gov
   F. Local Emergency Planning Committees. www.epa.gov/ceppo/epclist.htm

OSHA AND EMERGENCY RESPONSE

Even though no single OSHA standard is dedicated specifically to the issue of planning for chemical emergencies, all OSHA standards are written for the purpose of promoting a safe, healthy, accident-free, and hence emergency-free workplace. Therefore, OSHA standards do play a role in emergency prevention.

OSHA's standards, therefore, should be considered when developing emergency plans. A first step when developing emergency response plans is to review the guidelines presented earlier and to apply any and all applicable OSHA standards to your emergency response plan. This review of applicable guidelines and pertinent OSHA standards can help the managers of chemical industrial facilities identify and then correct conditions that might exacerbate emergency situations before they occur.

Conception of an Emergency Response Plan

Typically, when we think of emergency response plans for the workplace, we often conjure up thoughts about the obvious. For example, the first workplace emergency that might come to mind is fire—a major concern because fire in the workplace is something that can happen, that happens more often than we might think, and because fire can be particularly devastating—in ways we know all too well. Most employees do not need to be informed about the dangers of fire. However, employers have the responsibility to do just this—to inform and train employees on fire, fire prevention, and fire protection. Many local codes go beyond this information requirement, insisting that employers develop and implement a fire emergency response and/or evacuation plan. The primary emphasis has been on the latter—evacuation. However, if the employer equips a workplace with fire extinguishers and other firefighting equipment, and expects the employees to respond aggressively to extinguish workplace fires, then not only must the facility have an emergency response plan, the employer must also ensure that all company personnel called upon to fight the fire are completely trained on how to do so safely; this is stipulated in OSHA's 29 CFR 1910.156(c).157(g).

Another commonly considered workplace emergency response plan or scenario is designed and implemented for medical emergencies. Many chemical facilities satisfy this requirement simply by directing their employees to call 911 or some other emergency number whenever a medical emergency occurs in the workplace. Other facilities, though, may require employees to provide emergency first aid. When the employer chooses the employee-supplied first aid option, certain requirements must be met before any employee can legally administer first aid. First, the first aid responder must be trained and certified to administer first aid. This training aspect must also include training on OSHA's Bloodborne Pathogens Standard. This standard requires that the employee be trained on the dangers inherent in handling and being exposed to human body fluids. The employee must also be trained on how to protect himself or herself from body fluid contamination. If the first aid responder or anyone else is exposed to and contaminated by body fluids, the employer must make available the hepatitis B vaccine and vaccination series to all employees who have occupational exposure, and post-exposure evaluation and follow-up to all employees who have had an exposure incident; this requirement is stipulated in OSHA's 29 CFR 1910.1030.

A third type of emergency response plan required for implementation in selected (covered) facilities is OSHA's 29 CFR 1910.120 (Hazardous Waste Operations and Emergency Response—HAZWOPER) for releases of hazardous materials. Unless the facility operator can demonstrate that the operation does not involve employee exposure or the reasonable possibility for employee exposure to safety or health hazards, the following operations are covered:

1. Cleanup operations required by a governmental body involving hazardous substances conducted at uncontrolled hazardous waste sites, state priority site lists, sites recommended by the EPA, NPL, and initial investigations of government-identified sites that are conducted before the presence or absence of a hazardous substance has been ascertained.
2. Corrective actions involving cleanup operations at sites covered by the Resource Conservation and Recovery Act of 1976 (RCRA).
3. Voluntary cleanup operations at sites recognized by federal, state, local, or other governmental bodies as uncontrolled hazardous waste sites.
4. Operations involving hazardous waste conducted at treatment, storage, disposal (TSD) facilities regulated by RCRA.
5. Emergency response operations for releases of, or substantial threats of releases of, hazardous substances without regard to the location of the hazard.

The final requirement impacts the largest number of facilities that meet the criteria requiring full compliance with 29 CFR 1910.120 HAZWOPER, because many such

DID YOU KNOW?

OSHA's 29 CFR 1910.151 (a), (b), and (c), Medical Services and First Aid Standard, requires employers to (1) ensure the ready availability of medical personnel for advice and consultation on matters of health; (2) in the absence of an infirmary, clinic, or hospital in near proximity to the workplace which is used for the treatment of all injured employees, a person or persons shall be adequately trained to render first aid and adequate first aid supplies shall be readily available; and (3) where the eyes or body of any person may be exposed to injurious corrosive materials, suitable facilities for quick drenching of or flushing of the eyes and body shall be provided within the work area for immediate emergency use.

In auditing various industrial workplaces, we found that this particular OSHA standard is one of the most misunderstood by employers. Based on our experience, we found that OSHA's use of the words "near proximity to" and "adequate" contribute to the vagary and ambiguity of this standard. Most employers understand the need to provide first aid supplies in the workplace, but generally think that this is the extent of their responsibilities in this regard. Moreover, many workplaces do train their personnel on basic first aid and CPR, but do not require (in written job descriptions) that the trained employees respond to a workplace medical emergency.

We found that employers are reluctant to have employees respond to workplace medical emergencies because most employees do not want to respond. They do not want to be exposed to gore and the victim's pain. They also do not want to assume liability for trying to aid an injured victim.

The common response we received from workers can be summed up as follows: "When it comes to liability matters in America, let me sue you before you sue me."

facilities do not normally handle, store, treat, or dispose of hazardous waste, but do use or produce hazardous materials in their processes.

A good example of this type of facility is a wastewater treatment plant. A wastewater treatment plant, obviously, is designed to treat wastewater and its by-products (wastewater is not normally thought of as a hazardous material). However, common industry practice uses hazardous materials in the treatment of wastewater. Chlorine is one example. Sulfur dioxide, sodium hydroxide, anhydrous ammonia, and other hazardous substances are also commonly used. Because the use of hazardous materials could lead to an emergency from the release or spill of such materials, facilities using these materials must develop and employ an effective site emergency response plan.

Before we discuss the basic goals of an effective emergency response plan from an OSHA compliance point of view, we should define *emergency response*. Considering that individual facilities are different, with different dangers and different needs, defining emergency response is not always easy. However, for our purposes, we use the definition provided by CoVan (1995, 54):

> Emergency response is defined as a limited response to abnormal conditions expected to result in unacceptable risk requiring rapid corrective action to prevent harm to personnel, property, or system function.

CoVan (1995, 54) makes another important point about emergency response, one critical for the site manager and/or safety professional. He points out that "although emergency response and engineering tends toward prevention, emergency response is a skill area that safety engineers must be familiar with both because of regulations and good engineering practice." Good engineering practice is the law by which all competent managers and safety professionals work and live.

Now that we have defined emergency response, let's move on to the basic goals of an effective emergency response plan. Most of the currently available literature on this topic generally lists the goals as twofold: (1) minimize injury to facility personnel and (2) minimize damage to the facility and then return to normal operation as soon as possible.

Obviously, these goals make a great deal of good sense. However, you may be wondering about the language used—a couple of key words: "facility personnel" and "damage to the facility." Remember that we are talking about OSHA requirements here. Under OSHA the primary emphasis is on protecting the workers—protecting the workers' health and safety is OSHA's only focus.

What about people who live off-site—the site's neighbors?

What about the environment?

These questions stress the point we emphasize here. Again, OSHA is not normally concerned about the environment, unless contamination of the environment (at the work site) might adversely impact the workers' safety and health. The neighbors? Again, OSHA's focus is the workers. One OSHA compliance officer explained to us that if the employer takes every necessary step to protect the employees from harm involving the use or production of hazardous materials, then the surrounding community should have little to fear.

This statement is puzzling to us. We asked the same OSHA compliance officer about those incidents beyond the control of the employer—about accidents that could not only put employees in harm's way, but also endanger the surrounding community. The answer? "Well, that's the EPA's bag—we only worry about the work site and the worker."

Fortunately, OSHA, in combination with the U.S. Environmental Protection Agency (USEPA), has taken steps to overcome this blatant shortcoming (we like to think of it as an oversight). As mentioned earlier, under OSHA's Process Safety Management (PSM) and EPA's Risk Management Planning (RMP) directive, chemical spills and other chemical accidents that could impact both the environment and the "neighbors" have now been properly addressed. What PSM and RMP really accomplish is changing the typical twofold goal of an effective emergency response plan to a threefold goal.

Let's point out that the accomplishment of these twofold or threefold goals or objectives is essential in emergency response. Accomplishing these goals or objectives requires an extensive planning effort prior to the emergency ("prior" being the key word, because the attempts to develop an emergency response plan when a disaster is occurring or after one has occurred is both futile and stupid). The site manager and/or safety professional must never forget that while hazards in any facility can be reduced, risk is an element of everyday existence and therefore cannot be totally eliminated. The manager/safety professional's goal must be to keep risk to an absolute minimum. Again, to accomplish this, planning is critical as well as essential, and it should be accomplished well in advance of an emergency, because program testing should be part of the overall program design.

We pointed out earlier that most emergency plans address fire, medical emergencies, and the accidental release or spills of hazardous materials. Note that the development of emergency response plans should also factor in other possible emergencies—natural disasters, floods, explosions, and/or weather-related events that could occur and certainly will occur. Now, emergency response to terrorist activity or threats must also be added to the list.

Site emergency response plans should include:

- assessment of risk
- chain of command for dealing with emergencies
- assessment of resources
- training
- incident command procedures
- site security
- public relations

Table 8.1. Site emergency response plan

Elements to Be Included in the Plan	
Emergency response notification	List of whom to call and information to pass on when an emergency occurs
Record of changes	Table of changes and dates for them
Table of contents/ introduction	The purpose, objective, scope, applications, policies, and assumptions for the plan
Emergency response operations	Details what actions must take place
Emergency assistance telephone numbers	A current list of people and agencies who may be needed in an emergency
Legal authority and responsibility	References the laws and regulations that provide the authority for the plan
Chain of command	Response organization structure and responsibilities
Disaster assistance and coordination	Where additional assistance may be obtained when the regular response organizations are overburdened
Procedures for changing or updating the plan	Details who makes changes and how they are made and implemented
Plan distribution	List of organizations and individuals who have been given a copy of the plan
Spill cleanup techniques	Detailed information about how response teams should handle cleanups
Cleanup/disposal resources	List of what is available, where it is obtained, and how much is available
Consultant resources	List of special facilities and personnel who may be valuable in a response
Technical library/references	List of libraries and other information sources that may be valuable for those preparing, updating, or implementing the plan
Hazards analysis	Details the kinds of emergencies that may be encountered, where they are likely to occur, what areas of the community may be affected, and the probability of occurrence
Documentation of spill events	The various incident and investigative reports on spills that have occurred
Hazardous materials information	Listing of hazardous materials, their properties, response data, and related information
Dry runs	Training exercises for testing the adequacy of the plan, training personnel, and introducing changes

Source: Federal Emergency Management Agency (1981); adapted from Brauer (1994).

The Federal Emergency Management Agency (FEMA), the U.S. Army Corps of Engineers, and several other agencies, as well as numerous publications, provide guidance on how to develop a site emergency response plan. Local agencies (such as fire departments, emergency planning commissions/agencies, HazMat teams, and local emergency planning committees [LEPCs]) also provide information on how to design a site plan. All of these agencies typically recommend that a site's plan contain the elements listed in table 8.1.

In the site manager/safety professional's effort to incorporate and manage a facility emergency response plan, and in the response itself, two elements mentioned earlier (security considerations and public relations) must be given special attention. If not handled correctly, the lack of effective security measures and/or improper public relations can turn an already disastrous incident into a megadisaster.

In planning security considerations, provision should be made to have a well-trained security team limit site access to only the people and the equipment that will assist in coping with and resolving the emergency.

Public relations (PR) can be a tricky enterprise. The person identified to interface with the media must have thorough knowledge of the site, process, and personnel involved. The PR person must also have access to the highest levels of site management. Otherwise, he or she will not be able to deal with the public/media effectively.

To a certain extent, the person who is ultimately responsible for giving the go-ahead to starting up in response to a perceived emergency is in a "damned if you do, damned if you don't" position. When Hurricane Isabel hit the East Coast in 2003 with 100-mile-per-hour-plus winds and torrential rains, and as the effects of the storm were felt hundreds of miles away from the actual storm center, we all recognized that only slight changes in conditions could have pushed Isabel out to sea—and all that reaction to the approaching emergency would have looked like overkill. But that's both the beauty and the frustration of emergency planning. By calling in the process early, you may run the risk of looking overly cautious or, heaven forbid, wimpy. But if the storm does hit, no one is going to remember that you made the right call—you just did your job. In general, chemical site managers are people who would rather err on the side of caution (hopefully)—in other words, on the side of safety. Once the storm hits, it's too late to do much good if you weren't already geared up to go.

THE BOTTOM LINE

Because industrial emergencies (in less-than-extreme conditions) can seriously affect the surrounding community and environment, and because poor planning and/or panic can only make a bad situation worse and can also lead to additional injury and death, your role as chemical site manager or site safety professional in emergency response is doubly important. A crisis out of hand can easily devastate a community—and your organization is (or should be) an active member of your community. By ensuring less-than-effective emergency response, chemical site managers endanger not only themselves and their organizations but also endanger their organization's community and standing as well.

REFERENCES AND RECOMMENDED READING

Brauer, R. L. 1994. *Safety and health for engineers.* New York: Van Nostrand Reinhold.

CoVan, J. 1995. *Safety engineering.* New York: Wiley.

Federal Emergency Management Agency (FEMA). 1981. *Planning guide and checklist for hazardous materials contingency plans.* FEMA-10. Washington, DC: Federal Emergency Management Agency.

Healy, R. J. 1969. *Emergency and disaster planning.* New York: Wiley.

Occupational Safety and Health Administration (OSHA). 1987. *Occupational safety and health standard 29 CFR 1910.120: Hazardous waste operations and emergency response.* www.osha.gov/pls/oshaweb/owadisp.show_document?p_table=STANDARDS&p_id=9765.

Smith, A. J. 1980. *Managing hazardous substances accidents.* New York: McGraw-Hill.

Spellman, F. R. 1997. *A guide to compliance for process safety management planning (PSM/RMP).* Lancaster, PA: Technomic Publishing.

U.S. Army Corps of Engineers. 1987. *Safety and health requirements manual.* Rev. ed. EM 385-1-1. Washington, DC: U.S. Army Corps of Engineers.

U.S. Environmental Protection Agency (USEPA). 2003. *Large Water System Emergency Response Plan Outline: Guidance to Assist Community Water Systems in Complying with the Public Health Security and Bioterrorism Preparedness and Response Act of 2002.* EPA 810-F-03-007. www.epa.gov/safewater/security (accessed June 2006).

———. 2002. *Water Utility Response, Recovery, and Remediation Guidance for Man-Made and/or Technological Emergencies.* Washington, DC: U.S. Environmental Protection Agency.

9

Security Techniques and Hardware

We have tried to talk about it and the words fail us and we fall back silent as the gaping void.
To say nothing is best.
Towers no longer are, the skyline empty.
Yes, even in the face of great courage displayed, we live on through the tragedy of the loss.
To say nothing is best.
Yes, we go through the motions of grief and will take part in a national ceremony at memorial plaza, eventually to rebuild that which crumbled to dust.
But words are no help.
Towers no longer are, the skyline empty.
To say nothing is best.

—*Frank R. Spellman*

Ideally, in a perfect world, all chemical facilities would be secured in a layered fashion (aka the barrier approach). Layered security systems are vital. Using the protection "in-depth" principle, requiring that an adversary defeat several protective barriers or security layers to accomplish its goal, chemical industry infrastructure can be made more secure. Protection in depth is a term commonly used by the military to describe security measures that reinforce one another, masking the defense mechanisms from the view of intruders, and allowing the defender time to respond to intrusion or attack.

A prime example of the use of the multibarrier approach to ensure security and safety is demonstrated by the practices of the bottled water industry. In the aftermath of 9/11 and the increased emphasis on homeland security, a shifted paradigm of national security and vulnerability awareness has emerged. Recall that in the immediate aftermath of the 9/11 tragedies, emergency responders and others responded quickly and worked to exhaustion. In addition to the emergency responders, bottled water companies responded immediately by donating several million bottles of water to the crews at the crash sites in New York, at the Pentagon, and in Pennsylvania. The International Bottled Water Association (2004, 2) reports that "within hours of the first attack, bottled water was delivered where it mattered most; to emergency personnel on the scene who required ample water to stay hydrated as they worked to rescue victims and clean up debris."

Bottled water companies continued to provide bottled water to responders and rescuers at the 9/11 sites throughout the post-event processes. These patriotic actions by the bottled water companies, however, beg the question: How do we ensure the safety and security of the bottled water provided to anyone? IBWA (2004) has the answer: Using a multibarrier approach, along with other principles, will enhance the safety and security of bottled water. IBWA (2004, 3) describes its multibarrier approach as follows:

> Bottled water products are produced utilizing a multi-barrier approach, from source to finished product, that helps prevent possible harmful contaminants (physical, chemical or microbiological) from adulterating the finished product as well as storage, production, and transportation equipment. Measures in a multi-barrier approach may include source protection, source monitoring, reverse osmosis, distillation, filtration, ozonation or ultraviolet (UV) light. Many of the steps in a multi-barrier system may be effective in safeguarding bottled water from microbiological and other contamination. Piping in and out of plants, as well as storage silos and water tankers are also protected and maintained through sanitation procedures. In addition, bottled water products are bottled in a controlled, sanitary environment to prevent contamination during the filling operation.

In chemical industry infrastructure security, *protection in depth* is used to describe a layered security approach. A protection-in-depth strategy uses several forms of security techniques and/or devices against an intruder and does not rely on one single defensive mechanism to protect infrastructure. By implementing multiple layers of security, a hole or flaw in one layer is covered by the other layers. An intruder will have to intrude through each layer without being detected in the process—the layered approach implies that no matter how an intruder attempts to accomplish his goal, he will encounter effective elements of the physical protection system.

For example, as depicted in figure 9.1, an effective security layering approach requires that an adversary penetrate multiple, separate barriers to gain entry to a critical target at a chemical industry facility. As shown in figure 9.1, protection in depth (multiple layers of security) helps to ensure that the security system remains effective in the event of a failure or an intruder bypassing a single layer of security.

Again, as shown in figure 9.1, layered security starts with the outer perimeter (the fence—the first line of physical security) of the facility and goes inward to the facility, the buildings, structures, other individual assets, and finally to the contents of those buildings—the targets.

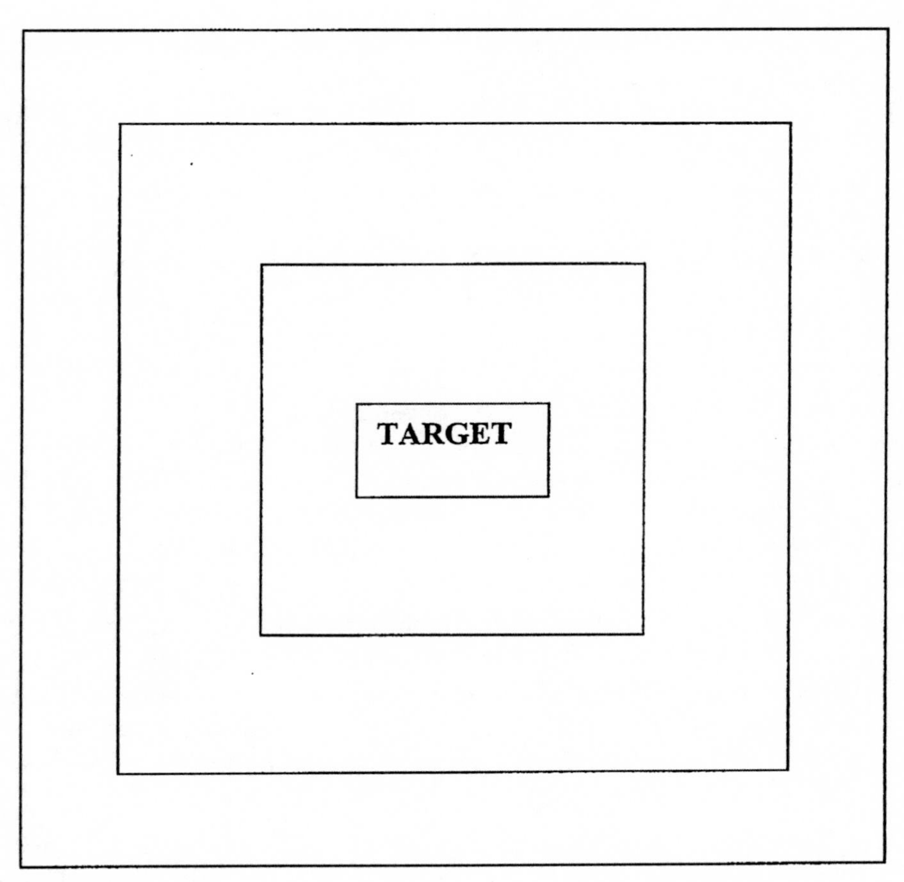

FIGURE 9.1
Layered approach to security.

The area between the outer perimeter and structures or buildings is known as the site. This open site area provides an incomparable opportunity for early identification of an unauthorized intruder and initiation of early warning/response. This open space area is commonly used to calculate the standoff distance; that is, it is the distance between the outside perimeter (public areas to the fence) and the target or critical assets (buildings/structures) inside the perimeter (inside the fence line—the restricted access area).

The open area, between the perimeter fence and the target (e.g., operations center), if properly outfitted with various security devices, can also provide layered protection against intruders. For example, lighting is a deterrent. Based on personal experience, an open area within the plant site that is almost as well lighted at night as would be expected during daylight hours is the rule of thumb. In addition, strategically placed motion detectors along with crash barriers at perimeter gate openings and in front of vital structures are also recommended. Armed, mobile guards who roam the interior of the plant site on a regular basis provide the ultimate in site area security.

The next layer of physical security is the outside wall of the target structure(s) itself. Notwithstanding door, window, and/or skylight entry, walls prevent most intruders from easy entry. If doors can only be entered using card reader access, security is shored up or enhanced to an extent. The same can be said for windows and skylights that are fashioned small enough to prohibit normal human entry. These same "weak" spots in buildings can be bastioned with break-proof or reinforced security glass.

The final layer of security is provided by properly designed interior features of buildings. Examples of these types of features include internal doors and walls, equipment cages, and backup or redundant equipment.

In the discussion above, conditions described referred to "perfect world" conditions—that is, to those conditions that we would want (i.e., the security manager's proverbial wish list) to be incorporated into the design and installation of new chemical industry infrastructure. Post-9/11, in a not-so-perfect world, however, many of the peripheral (fence line) measures described above are more difficult to incorporate into chemical industry site infrastructure. This is not to say that industrial chemical facilities do not have fence lines or fences; most of them do. These fences are designed to keep vandals, thieves, and trespassers out. The problem is that many of these facilities were constructed several years ago, before urban encroachment literally encircled the sites—allowing, at present, little room for security stand-backs or setbacks to be incorporated into plants or critical equipment locations. Based on personal observation, many of these fences face busy city streets or closely abut structures outside the fence line. The point is that when one sits down to plan a security upgrade, these factors must be taken into account.

For existing facilities, security upgrades should be based on the results generated from the vulnerability assessment, which characterizes and prioritizes those assets that may be targeted. The vulnerabilities identified must be protected.

In the following sections, various security hardware and/or devices are described. These devices serve the main purpose of providing security against physical and/or digital intrusion. That is, they are designed to delay and deny intrusion and are normally coupled with detection and assessment technology. However, as mentioned previously, no matter the type of security device or system employed, chemical industry systems cannot be made immune to all possible intrusions. Simply, when it comes to making anything absolutely secure from intrusion or attack, there is, inherently or otherwise, no silver bullet.

USEPA (2005) groups infrastructure security devices or products described below into four general categories:

- physical asset monitoring and control devices
- cyber protection devices
- communication/integration
- environmental monitoring devices (not covered; beyond context of text)

PHYSICAL ASSET MONITORING AND CONTROL DEVICES

Aboveground, Outdoor Equipment Enclosures

Chemical industrial complexes usually consist of multiple components spread over a wide area, and typically include a centralized chemical processing plant, as well as chemical storage and distribution system components that are typically distributed at multiple locations throughout the plant site. In recent years, chemical plant system designers have favored placing critical equipment—especially assets that require regular use and maintenance—aboveground. One of the primary reasons for doing so is that locating this equipment aboveground eliminates the safety risks associated with confined space entry, which is often required for the maintenance of equipment located belowground. In addition, space restrictions often limit the amount of equipment that can be located inside, and there are concerns that some types of equipment (such as backflow prevention devices—to prevent chemicals from entering plant and off-site potable water systems) can, under certain circumstances, discharge chemical slurry mixtures that could flood pits, vaults, or equipment rooms. Therefore, many pieces of critical equipment are located outdoors and aboveground.

Experience demonstrates that many different system components can be and are often installed outdoors and aboveground. Examples of these types of components include the following:

- backflow prevention devices
- air release and control valves
- pressure vacuum breakers
- pumps and motors
- chemical storage and feed equipment
- meters
- sampling equipment
- instrumentation

Much of this equipment is installed in remote locations and/or in areas where the public (and terrorists) can access it.

One of the most effective security measures for protecting aboveground equipment is to place it inside a building. When/where this is not possible, enclosing the equipment or parts of the equipment using some sort of commercial or homemade add-on structure may help to prevent tampering with the equipment. These types of add-on structures or enclosures, which are designed to protect the equipment both from the elements and from unauthorized access or tampering, typically consist of a box-like structure that is placed over the entire component, or over critical parts of the component (i.e., valves, etc.), and is then secured to delay or prevent intruders from tampering with the equipment. The enclosures are typically locked or otherwise anchored to a solid foundation, which makes it difficult for unauthorized personnel to remove the enclosure and access the equipment.

Standardized aboveground enclosures are available in a wide variety of materials, sizes, and configurations. Many options and security features are also available for each type of enclosure, and this allows system operators the flexibility to customize an enclosure for a specific application and/or price range. In addition, most manufacturers can custom-design enclosures if standard, off-the-shelf enclosures do not meet a user's needs.

Many of these enclosures are designed to meet certain standards. For example, the American Society of Sanitary Engineers (ASSE) has developed Standard 1060, *Performance Requirements for Outdoor Enclosures for Backflow Prevention Assemblies.* If an enclosure will be used to house a backflow preventer, this standard specifies the acceptable construction materials for the enclosure, as well as the performance requirements that the enclosure should meet, including specifications for freeze protection, drainage, air inlets, access for maintenance, and hinge requirements. ASSE 1060 also states that the enclosure should be lockable to enhance security.

Equipment enclosures can generally be categorized into one of four main configurations, which include

- one-piece, drop-over enclosures
- hinged- or removable-top enclosures
- sectional enclosures
- shelters with access locks

All enclosures, including those with integral floors, must be secured to a foundation to prevent them from being moved or removed. Unlocked or poorly-anchored enclosures may be blown off the equipment being protected, or may be defeated by intruders. In either case, this may result in the equipment beneath the enclosure becoming exposed and damaged. Therefore, ensuring that the enclosure is securely anchored will increase the security of the protected equipment.

The three basic types of foundations that can be used to anchor the aboveground equipment enclosure are concrete footers, concrete slabs-on-grade, or manufactured fiberglass pads. The most common types of foundations utilized for equipment enclosures are standard or slab-on-grade footers; however, local climate and soil conditions may dictate whether either of these types of foundations can be used. These foundations can be either precast or poured in place at the installation site. Once the foundation is installed and properly cured, the equipment enclosure is bolted or anchored to the foundation to secure it in place.

An alternative foundation, specifically for use with smaller hot box enclosures, is a manufactured fiberglass pad known as the Glass Pad™. The Glass Pad has the center cut out so that it can be dropped directly over the piece of equipment being enclosed. Once the pad is set level on the ground, it is backfilled over a two-inch flange located around its base. The enclosure is then placed on top of the foundation, and is locked in place with either a staple- or a slotted-anchor, depending on the enclosure configuration.

One of the primary attributes of a security enclosure is its strength and resistance to breaking and penetration. Accordingly, the materials from which the enclosure is constructed will be important in determining the strength of the enclosure, and thus its usefulness for security applications. Enclosures are typically manufactured from either fiberglass or aluminum. With the exception of the one-piece, drop-over enclosure, which is typically fabricated from fiberglass, each configuration described above can be constructed from either material. In addition, enclosures can be custom-manufactured from polyurethane, galvanized steel, or stainless steel. Galvanized or stainless steel is often offered as an exterior layer, or "skin," for an aluminum enclosure. Although they are typically utilized in underground applications, precast concrete structures can also be used as aboveground equipment enclosures. However, precast structures are much heavier and more difficult to maneuver than are their fiberglass and aluminum counterparts. Concrete is also brittle, and that can be a security concern; however, products can be applied to concrete structures to add strength and minimize security risks (i.e., epoxy coating). Because precast concrete structures can be purchased from any concrete producers, this document does not identify specific vendors for these types of products.

In addition to the construction materials, enclosure walls can be configured or reinforced to give them added strength. Adding insulation is one option that can strengthen the structural characteristics of an enclosure; however, some manufacturers offer additional features to add strength to exterior walls. For example, while most enclosures are fabricated with a flat wall construction, some vendors manufacture fiberglass shelters with ribbed exterior walls. These ribs increase the structural integrity of the wall and allow the fabrication of standard shelters up to twenty feet in length. Another vendor has developed a proprietary process that uses a series of integrated fiberglass beams that are placed throughout a foam inner core to tie together the interior and exterior walls and roof. Yet another vendor constructs aluminum enclosures with horizontal and vertical redwood beams for structural support.

Other security features that can be implemented on aboveground, outdoor equipment enclosures include locks, mounting brackets, tamper-resistant doors, and exterior lighting.

Active Security Barriers (Crash Barriers)

Active security barriers (also known as crash barriers) are large structures that are placed in roadways at entrance and exit points to protected facilities to control vehicle access to these areas. These barriers are placed perpendicular to traffic to block the roadway, so that the only way that traffic can pass the barrier is for the barrier to be moved out of the roadway. These types of barriers are typically constructed from sturdy materials, such as concrete or steel, such that vehicles cannot penetrate them. They are also designed at a certain height off the roadway so that vehicles cannot go over them.

The key difference between active security barriers (which include wedges, crash beams, gates, retractable bollards, and portable barricades) and passive security barriers (which include non-movable bollards, jersey barriers, and planters) is that active security barriers are designed so that they can be raised and lowered or moved out of the roadway easily to allow authorized vehicles to pass. Many of these types of barriers are designed so that they can be opened and closed automatically (i.e., mechanized gates, hydraulic wedge barriers), while others are easy to open and close manually (swing crash beams, manual gates). In contrast to active barriers, passive barriers are permanent, non-movable barriers, and thus they are typically used to protect the perimeter of a protected facility, such as sidewalks and other areas that do not require vehicular traffic to pass them. Several of the major types of active security barriers (wedge barriers, crash beams, gates, bollards, and portable/removable barricades) are described below.

Wedge barriers are plated, rectangular steel buttresses approximately two to three feet high that can be raised and lowered from the roadway. When they are in the open position, they are flush with the roadway and vehicles can pass over them. However, when they are in the closed (armed) position, they project up from the road at a forty-five-degree angle, with the upper end pointing toward the oncoming vehicle and the base of the barrier away from the vehicle. Generally, wedge barriers are constructed from heavy-gauge steel, or concrete that contains an impact-dampening iron rebar core that is strong and resistant to breaking or cracking, thereby allowing them to withstand the impact from a vehicle attempting to crash through them. In addition, both of these materials help to transfer the energy of the impact over the barrier's entire volume, thus helping to prevent the barrier from being sheared off its base.

In addition, because the barrier is angled away from traffic, the force of any vehicle impacting the barrier is distributed over the entire surface of the barrier and is not concentrated at the base, which helps prevent the barrier from breaking off at the base. Finally, the angle of the barrier helps hang up any vehicle attempting to drive over it.

Wedge barriers can be fixed or portable. Fixed wedge barriers can be mounted on the surface of the roadway (surface-mounted wedges) or in a shallow mount in the road's surface, or they can be installed completely below the road surface. Surface-mounted wedge barricades operate by rising from a flat position on the surface of the roadway, while shallow-mounted wedge barriers rise from their resting position just below the road surface. In contrast, below-surface wedge barriers operate by rising from beneath the road surface. Both the shallow-mounted and surface-mounted barriers require little or no excavation, and thus do not interfere with buried utilities. All three barrier mounting types project above the road surface and block traffic when they are raised into the armed position. Once they are disarmed and lowered, they are flush with the road, thereby allowing traffic to pass. Portable wedge barriers (see below) are moved into place on wheels that are removed after the barrier has been set into place.

Installing rising wedge barriers requires preparation of the road surface. Installing surface-mounted wedges does not require that the road be excavated; however, the road surface must be intact and strong enough to allow the bolts anchoring the wedge to the road surface to attach properly. Shallow-mount and below-surface wedge barricades require excavation of a pit that is large enough to accommodate the wedge structure, as well as any arming/disarming mechanisms. Generally, the bottom of the excavation pit is lined with gravel to allow for drainage. Areas not sheltered from rain or surface runoff can have installed a gravity drain or self-priming pump. Table 9.1 lists the pros and cons of wedge barriers.

Crash beam barriers consist of aluminum beams that can be opened or closed across the roadway. While there are several different crash beam designs, every crash beam system consists of an aluminum beam that is supported on each side by a solid footing or buttress, which is typically constructed from concrete, steel, or some other strong material. Beams typically contain an interior steel cable (usually at least one inch in diameter) to give the beam added strength and rigidity. The beam is connected by a heavy-duty hinge or other mechanism to one of the footings so that it can swing or rotate out of the roadway when it is open and can swing back across the road when it is in the closed (armed) position, blocking the road and inhibiting access by unauthorized vehicles. The non-hinged end of the beam can be locked into its footing, thus providing anchoring for the beam on both sides of the road and increasing the beam's resistance to any vehicles attempting to penetrate through it. In addition, if the crash beam is hit by a vehicle, the aluminum beam transfers the impact energy to the interior

Table 9.1. Pros and cons of wedge barriers

Pros	*Cons*
Can be surface-mounted or completely installed below the roadway surface.	Installations below the surface of the roadway will require construction that may interfere with buried utilities.
Wedge barriers have a quick response time (normally 3.5–10.5 seconds, but can be 1–3 seconds in emergency situations. Because emergency activation of the barrier causes more wear and tear on the system than does normal activation, it is recommended for use only in true emergency situations.	Regular maintenance is needed to keep wedge fully operational.
Surface- or shallow-mount wedge barricades can be utilized in locations with a high water table and/or corrosive soils.	Improper use of the system may result in authorized vehicles being hung up by the barrier and damaged. Guards must be trained to use the system properly to ensure that this does not happen. Safety technologies may also be installed to reduce the risk of the wedge activating under an authorized vehicle.
All three wedge barrier designs have a high crash rating, thereby allowing them to be employed for higher-security applications.	
These types of barrier are extremely visible, which may deter potential intruders.	

Source: USEPA (2005).

cable, which in turn transfers the impact energy through the footings and into their foundation, thereby minimizing the chance that the impact will snap the beam and allow the intruding vehicle to pass through.

Crash beam barriers can employ drop-arm, cantilever, or swing beam designs. Drop-arm crash beams operate by raising and lowering the beam vertically across the road. Cantilever crash beams are projecting structures that are opened and closed by extending the beam from the hinge buttress to the receiving buttress located on the opposite side of the road. In the swing beam design, the beam is hinged to the buttress so that it swings horizontally across the road. Generally, swing beam and cantilever designs are used at locations where a vertical lift beam is impractical. For example, the swing beam or cantilever design is utilized at entrances and exits with overhangs, trees, or buildings that would physically block the operation of the drop-arm beam design.

Installing any of these crash beam barriers involves the excavation of a pit approximately forty-eight inches deep for both the hinge and the receiver footings. Due to the depth of excavation, the site should be inspected for underground utilities before digging begins. Table 9.2 lists the pros and cons of crash beams.

In contrast to wedge barriers and crash beams, which are typically installed separately from a fence line, *gates* are often integrated units of a perimeter fence or wall around a facility. Gates are basically movable pieces of fencing that can be opened and

Table 9.2. Pros and cons of crash beams

Pros	*Cons*
Requires little maintenance, while providing long-term durability.	Crash beams have a slower response time (normally 9.5–15.3 seconds, but can be reduced to 7–10 seconds in emergency situations) than do other types of active security barriers, such as wedge barriers. Because emergency activation of the barrier causes more wear and tear on the system than does normal activation, it is recommended for use only in true emergency situations.
No excavation is required in the roadway itself to install crash beams.	All three crash beam designs possess a low crash rating relative to other types of barriers, such as wedge barriers, and thus they typically are used for lower-security applications.
	Certain crash barriers may not be visible to oncoming traffic and therefore may require additional lighting and/or other warning markings to reduce the potential for traffic to accidentally run into the beam.

Source: USEPA (2005).

closed across a road. When the gate is in the closed (armed) position, the leaves of the gate lock into steel buttresses that are embedded in concrete foundations located on both sides of the roadway, thereby blocking access to the roadway. Generally, gate barricades are constructed from a combination of heavy-gauge steel and aluminum that can absorb an impact from vehicles attempting to ram through them. Any remaining impact energy not absorbed by the gate material is transferred to the steel buttresses and their concrete foundation.

Gates can utilize a cantilever, linear, or swing design. Cantilever gates are projecting structures that operate by extending the gate from the hinge footing across the roadway to the receiver footing. A linear gate is designed to slide across the road on tracks via a rack-and-pinion drive mechanism. Swing gates are hinged so that they can swing horizontally across the road.

Installation of the cantilever, linear, or swing gate designs described above involve the excavation of a pit approximately forty-eight inches deep for both the hinge and receiver footings to which the gates are attached. Due to the depth of excavation, the site should be inspected for underground utilities before digging begins. Table 9.3 lists the pros and cons of gates.

Bollards are vertical barriers at least three feet tall and one to two feet in diameter that are typically set four to five feet apart from each other so that they block vehicles from passing between them. Bollards can either be fixed in place, removable, or retractable.

Table 9.3. Pros and cons of gates

Pros	*Cons*
All three gate designs possess an intermediate crash rating, thereby allowing them to be utilized for medium- to higher-security applications.	Gates have a slower response time (normally 10–15 seconds, but can be reduced to 7–10 seconds in emergency situations) than do other types of active security barriers, such as wedge barriers. Because emergency activation of the barrier causes more wear and tear on the system than does normal activation, it is recommended for use only in true emergency situations.
Requires very little maintenance.	
Can be tailored to blend in with perimeter fencing.	
Gate construction requires no roadway excavation.	
Cantilever gates are useful for roads with high crowns or drainage gutters.	
These types of barriers are extremely visible, which may deter intruders.	
Gates can also be used to control pedestrian traffic.	

Source: USEPA (2005).

Fixed and removable bollards are passive barriers that are typically used along building perimeters or on sidewalks to prevent vehicles from passing through, while allowing pedestrians to pass between them. In contrast to passive bollards, retractable bollards are active security barriers that can easily be raised and lowered to allow vehicles to pass between them. Thus, they can be used in driveways or on roads to control vehicular access. When the bollards are raised, they project above the road surface and block the roadway; when they are lowered, they sit flush with the road surface and thus allow traffic to pass over them. Retractable bollards are typically constructed from steel or other materials that have a low weight-to-volume ratio so that they require low power to raise and lower. Steel is also more resistant to breaking than is a more brittle material, such as concrete, and is better able to withstand direct vehicular impact without breaking apart.

Retractable bollards are installed in a trench dug across a roadway—typically at an entrance or gate. Installing retractable bollards requires preparing the road surface. Depending on the vendor, bollards can be installed either in a continuous slab of concrete, or in individual excavations with concrete poured in place. The required excavation for a bollard is typically slightly wider and slightly deeper than the bollard height when extended aboveground. The bottom of the excavation is typically lined with gravel to allow drainage. The bollards are then connected to a control panel which controls the raising and lowering of the bollards. Installation typically requires

Table 9.4. Pros and cons of retractable bollards

Pros	*Cons*
Bollards have a quick response time (normally 3–10 seconds, but can be reduced to 1–3 seconds in emergency situations).	Bollard installations will require construction below the surface of the roadway, which may interfere with buried utilities.
Bollards have an intermediate crash rating, which allows them to be utilized for medium- to higher-security applications.	Some maintenance is needed to ensure barrier is free to move up and down.
	The distance between bollards must be decreased (i.e., more bollards must be installed along the same perimeter) to make these systems effective against small vehicles (i.e., motorcycles).

Source: USEPA (2005).

mechanical, electrical, and concrete work; if utility personnel with these skills are available, then the utility can install the bollards themselves. Table 9.4 lists the pros and cons of retractable bollards.

Portable/removable barriers, which can include removable crash beams and wedge barriers, are mobile obstacles that can be moved in and out of position on a roadway. For example, a crash beam may be completely removed and stored off-site when it is not needed. An additional example would be wedge barriers that are equipped with wheels that can be removed after the barricade is towed into place.

When portable barricades are needed, they can be moved into position rapidly. To provide them with added strength and stability, they are typically anchored to buttress boxes that are located on either side of the road. These buttress boxes, which may or may not be permanent, are usually filled with sand, water, cement, gravel, or concrete to make them heavy and aid in stabilizing the portable barrier. In addition, these buttresses can help dissipate any impact energy from vehicles crashing into the barrier itself.

Because these barriers are not anchored into the roadway, they do not require excavation or other related construction for installation. In contrast, they can be assembled and made operational in a short period of time. The primary shortcoming to this type of design is that these barriers may move if they are hit by vehicles. Therefore, it is important to carefully assess the placement and anchoring of these types of barriers to ensure that they can withstand the types of impacts that may be anticipated at that location. Table 9.5 lists the pros and cons of portable/removable barricades.

Because the primary threat to active security barriers is that vehicles will attempt to crash through them, their most important attributes are their size, strength, and crash resistance. Other important features for an active security barrier are the mechanisms by which the barrier is raised and lowered to allow authorized vehicle entry, and other factors such as weather resistance and safety features.

Table 9.5. Pros and cons of portable/removable barricades

Pros	*Cons*
Installing portable barricades requires no foundation or roadway excavation.	Portable barriers may move slightly when hit by a vehicle, resulting in a lower crash resistance.
Can be moved in and out of position in a short period of time.	Portable barricades typically require 7.75 to 16.25 seconds to move into place, and thus they are considered to have a medium response time when compared with other active barriers.
Wedge barriers equipped with wheels can be easily towed into place.	
Minimal maintenance is needed to keep barriers fully operational.	

Source: USEPA (2005).

Alarms

An *alarm system* is a type of electronic monitoring system that is used to detect and respond to specific types of events—such as unauthorized access to an asset, or a possible fire. In chemical processing systems, alarms are also used to alert operators when process operating or monitoring conditions go out of preset parameters (i.e., process alarms). These types of alarms are primarily integrated with process monitoring and reporting systems (i.e., SCADA systems). Note that this discussion does not focus on alarm systems that are not related to a facility's processes.

Alarm systems can be integrated with fire detection systems, intrusion detection systems (IDSs), access control systems, or closed circuit television (CCTV) systems, so that these systems automatically respond when the alarm is triggered. For example, a smoke detector alarm can be set up to automatically notify the fire department when smoke is detected, or an intrusion alarm can automatically trigger cameras to turn on in a remote location so that personnel can monitor that location.

An alarm system consists of sensors that detect different types of events; an arming station that is used to turn the system on and off; a control panel that receives information, processes it, and transmits the alarm; and an annunciator that generates a visual and/or audible response to the alarm. When a sensor is tripped it sends a signal to a control panel, which triggers a visual or audible alarm and/or notifies a central monitoring station. A more complete description of each of the components of an alarm system is provided below.

Detection devices (also called *sensors*) are designed to detect a specific type of event (such as smoke, intrusion, etc.). Depending on the type of event they are designed to detect, sensors can be located inside or outside of the facility or other asset. When an event is detected, the sensors use some type of communication method (such as wireless radio transmitters, conductors, or cables) to send signals to the control panel to generate the alarm. For example, a smoke detector sends a signal to a control panel when it detects smoke.

Alarms use either normally closed (NC) or normally open (NO) electric loops, or "circuits," to generate alarm signals. In NC loops or circuits, all of the system's sensors and switches are connected in series. The contacts are "at rest" in the closed (on) position, and current continually passes through the system. However, when an event triggers the sensor, the loop is opened, breaking the flow of current through the system and triggering the alarm. NC switches are used more often than are NO switches because the alarm will be activated if the loop or circuit is broken or cut, thereby reducing the potential for circumventing the alarm. This is known as a "supervised" system.

In NO loops or circuits, all of the system's sensors and switches are connected in parallel. The contacts are "at rest" in the open (off) position, and no current passes through the system. However, when an event triggers the sensor, the loop is closed. This allows current to flow through the loop, powering the alarm. NO systems are not "supervised" because the alarm will not be activated if the loop or circuit is broken or cut. However, adding an end-of-line resistor to an NO loop will cause the system to alarm if tampering is detected.

An *arming station*, which is the main user interface with the security system, allows the user to arm (turn on), disarm (turn off), and communicate with the system. How a specific system is armed will depend on how it is used. For example, while IDSs can be armed for continuous operation (twenty-four hours/day), they are usually armed and disarmed according to the work schedule at a specific location so that personnel going about their daily activities do not set off the alarms. In contrast, fire protection systems are typically armed twenty-four hours/day.

A *control panel* receives information from the sensors and sends it to an appropriate location, such as to a central operations station or to a twenty-four-hour monitoring facility. Once the alarm signal is received at the central monitoring location, personnel monitoring for alarms can respond (such as by sending security teams to investigate or by dispatching the fire department).

An *annunciator* responds to the detection of an event by emitting a signal. This signal may be visual, audible, electronic, or a combination of these three. For example, fire alarm signals will always be connected to audible annunciators, whereas intrusion alarms may not be.

Alarms can be reported locally, remotely, or both locally and remotely. A *local alarm* emits a signal at the location of the event (typically using a bell or siren). A "local only" alarm emits a signal at the location of the event but does not transmit the alarm signal to any other location (i.e., it does not transmit the alarm to a central monitoring location). Typically, the purpose of a "local only" alarm is to frighten away intruders, and possibly to attract the attention of someone who might notify the proper authorities. Because no signal is sent to a central monitoring location, personnel can only respond to a local alarm if they are in the area and can hear and/or see the alarm signal.

Fire alarm systems must have local alarms, including both audible and visual signals. Most fire alarm signal and response requirements are codified in the National Fire Alarm Code, National Fire Protection Association (NFPA) 72. NFPA 72 discusses the application, installation, performance, and maintenance of protective signaling systems and their components. In contrast to fire alarms, which require a local signal when fire is detected, many IDSs do not have a local alert device, because monitoring personnel do not wish to inform potential intruders that they have been detected. Instead, these types of systems silently alert monitoring personnel that an intrusion has been detected, thus allowing monitoring personnel to respond.

In contrast to systems that are set up to transmit "local only" alarms when the sensors are triggered, systems can also be set up to transmit signals to a *central location*, such as to a control room or guard post at the utility, or to a police or fire station. Most fire/smoke alarms are set up to signal both at the location of the event and at a fire station or central monitoring station. Many insurance companies require that facilities install certified systems that include alarm communication to a central station. For example, systems certified by the Underwriters Laboratory (UL) require that the alarm be reported to a central monitoring station.

The main differences between alarm systems lie in the types of event detection devices used in different systems. *Intrusion sensors*, for example, consist of two main categories: perimeter sensors and interior (space) sensors. *Perimeter intrusion sensors* are typically applied on fences, doors, walls, windows, etc., and are designed to detect intruders before they access a protected asset (i.e., perimeter intrusion sensors are used to detect intruders attempting to enter through a door, window, etc.). In contrast, *interior intrusion sensors* are designed to detect an intruder who has already accessed the protected asset (i.e., interior intrusion sensors are used to detect intruders once they are already within a protected room or building). These two types of detection devices can be complementary, and they are often used together to enhance security for an asset. For example, a typical intrusion alarm system might employ a perimeter glass-break detector that protects against intruders accessing a room through a window, as well as an ultrasonic interior sensor that detects intruders that have gotten into the room without using the window. Table 9.6 lists and describes types of perimeter and interior sensors.

Fire detection/fire alarm systems consist of different types of fire detection devices and fire alarm systems. These systems may detect fire, heat, smoke, or a combination of any of these. For example, a typical fire alarm system might consist of heat sensors, which are located throughout a facility and which detect high temperatures or a

Table 9.6. Perimeter and interior sensors

Type of Perimeter Sensor	*Description*
Foil	Foil is a thin, fragile, lead-based metallic tape that is applied to glass windows and doors. The tape is applied to the window or door, and electric wiring connects this tape to a control panel. The tape functions as a conductor and completes the electric circuit with the control panel. When an intruder breaks the door or window, the fragile foil breaks, opening the circuit and triggering an alarm condition.
Magnetic switches (reed switches)	The most widely used perimeter sensor. They are typically used to protect doors, as well as windows that can be opened (windows that cannot be opened are more typically protected by foil alarms).
Glass-break detectors	Placed on glass and sense vibrations in the glass when it is disturbed. The two most common types of glass-break detectors are shock sensors and audio discriminators.
Type of Interior Sensor	
Passive infrared (PIR)	Presently the most popular and cost-effective interior sensors. PIR detectors monitor infrared radiation (energy in the form of heat) and detect rapid changes in temperature within a protected area. Because infrared radiation is emitted by all living things, these types of sensors can be very effective.
Quad PIRs	Consist of two dual-element sensors combined in one housing. Each sensor has a separate lens and a separate processing circuitry, which allows each lens to be set up to generate a different protection pattern.
Ultrasonic detectors	Emit high-frequency sound waves, and sense movement in a protected area by sensing changes in these waves. The sensor emits sound waves that stabilize and set a baseline condition in the area to be protected. Any subsequent movement within the protected area by a would-be intruder will cause a change in these waves, thus creating an alarm condition.
Microwave detectors	Emit ultra-high-frequency radio waves, and the detector senses any changes in these waves as they are reflected throughout the protected space. Microwaves can penetrate through walls, and thus a unit placed in one location may be able to protect multiple rooms.
Dual technology devices	Incorporate two different types of sensor technology (such as PIR and microwave technology) together in one housing. When both technologies sense an intrusion, an alarm is triggered.

Source: USEPA (2005).

certain change in temperature over a fixed time period. A different system might be outfitted with both smoke and heat detection devices. A summary of several different types of fire/smoke/heat detection sensors is provided in table 9.7.

Once a sensor in an alarm system detects an event, it must communicate an alarm signal. The two basic types of alarm communication systems are hardwired and wireless. Hardwired systems rely on wire that is run from the control panel to each of the detection devices and annunciators. Wireless systems transmit signals from a transmitter to a receiver through the air—primarily using radio or other waves. Hardwired

Table 9.7. Fire/smoke/heat detection sensors

Detector Type	*Description*
Thermal detector	Senses when temperatures exceed a set threshold (fixed temperature detector) or when the rate of change of temperature increases over a fixed time period (rate-of-rise detector).
Duct detector	Is located within the heating and ventilation ducts of the facility. This sensor detects the presence of smoke within the system's return or supply ducts. A sampling tube can be added to the detector to help span the width of the duct.
Smoke detector	Senses invisible and/or visible products of combustion. The two principal types of smoke detector are photoelectric and ionization detectors. The major differences between these devices are described below: • Photoelectric smoke detectors react to visible particles of smoke. These detectors are more sensitive to the cooler smoke with large smoke particles that is typical of smoldering fires. • Ionization smoke detectors are sensitive to the presence of ions produced by the chemical reactions that take place with few smoke particles, such as those typically produced by fast-burning/flaming fires.
Multi-sensor detector	Is a combination of photoelectric and thermal detectors. The photoelectric sensor serves to detect smoldering fires, while the thermal detector senses the heat given off from fast-burning/flaming fires.
Carbon monoxide (CO) detector	Is used to indicate the outbreak of fire by sensing the level of carbon monoxide in the air. The detector has an electrochemical cell that senses carbon monoxide, but not other products of combustion.
Beam detector	Is designed to protect large, open spaces such as industrial warehouses. These detectors consist of three parts: the transmitter, which projects a beam of infrared light; the receiver, which registers the light and produces an electrical signal; and the interface, which processes the signal and generates alarm or fault signals. In the event of a fire, smoke particles obstruct the beam of light. Once a preset threshold is exceeded, the detector will go into alarm.
Flame detector	Senses either ultraviolet (UV) or infrared (IR) radiation emitted by a fire.
Air-sampling detector	Actively and continuously samples the air from a protected space and is able to sense the pre-combustion stages of incipient fire.

Source: USEPA (2005).

systems are usually lower-cost, more reliable (they are not affected by terrain or environmental factors), and significantly easier to troubleshoot than are wireless systems. However, a major disadvantage of hardwired systems is that it may not be possible to hardwire all locations (for example, it may be difficult to hardwire remote locations). In addition, running wires to their required locations can be both time consuming and costly. The major advantage to using wireless systems is that they can often be installed in areas where hardwired systems are not feasible. However, wireless components can be much more expensive when compared to hardwired systems. In addition, in the past, it has been difficult to perform self-diagnostics on wireless systems to confirm that they are communicating properly with the controller. Presently, the majority of wireless systems incorporate supervising circuitry, which allows the subscriber to know immediately if there is a problem with the system (such as a broken detection device or a low battery), or if a protected door or window has been left open.

Backflow Prevention Devices

Backflow prevention devices are designed to prevent backflow, which is the reversal of the normal and intended direction of water flow in a water system. Backflow is a potential problem in a chemical processing system because if incorrectly cross-connected to potable water it can spread contaminated water back through a distribution system. For example, backflow at uncontrolled cross connections (any actual or potential connection between the public water supply and a source of chemical contamination) can allow pollutants or contaminants to enter the potable water system. More specifically, backflow from private plumbing systems, industrial areas, hospitals, and other hazardous contaminant–containing systems, into public water mains and wells poses serious public health risks and security problems. Cross contamination from private plumbing systems can contain biological hazards (such as bacteria or viruses) or toxic substances that can contaminate and sicken an entire population in the event of backflow. The majority of historical incidences of backflow have been accidental, but growing concern that contaminants could be intentionally backfed into a system is prompting increased awareness for private homes, businesses, industries, and areas most vulnerable to intentional strikes. Therefore, backflow prevention is a major tool for the protection of water systems.

Backflow may occur under two types of conditions: backpressure and backsiphonage. *Backpressure* is the reverse from normal flow direction within a piping system that is the result of the downstream pressure being higher than the supply pressure. These reductions in the supply pressure occur whenever the amount of water being used exceeds the amount of water supplied, such as during water line flushing, firefighting, or breaks in water mains. *Backsiphonage* is the reverse from normal flow direction within a piping system that is caused by negative pressure in the supply piping (i.e., the reversal of normal flow in a system caused by a vacuum or partial vacuum within the water supply piping). Backsiphonage can occur where there is a high velocity in a pipe line; when there is a line repair or break that is lower than a service point; or when there is lowered main pressure due to high water withdrawal rate, such as during firefighting or water main flushing.

To prevent backflow, various types of backflow preventers are appropriate for use. The primary types of backflow preventers are:

- air gap drains
- double check valves
- reduced pressure principle assemblies
- pressure vacuum breakers

Biometric Security Systems

Biometrics involves measuring the unique physical characteristics or traits of the human body. Any aspect of the body that is measurably different from person to person—for example, fingerprints or eye characteristics—can serve as a unique biometric identifier for that individual. Biometric systems recognizing fingerprints, palm shape, eyes, face, voice, and signature comprise the bulk of the current biometric systems.

Biometric security systems use biometric technology combined with some type of locking mechanism to control access to specific assets. In order to access an asset controlled by a biometric security system, an individual's biometric trait must be matched with an existing profile stored in a database. If there is a match between the two, the locking mechanism (which could be a physical lock, such as at a doorway; an electronic lock, such as at a computer terminal; or some other type of lock) is disengaged, and the individual is given access to the asset.

A biometric security system is typically comprised of the following components:

- A sensor, which measures/records a biometric characteristic or trait.
- A control panel, which serves as the connection point between various system components. The control panel communicates information back and forth between the sensor and the host computer, and controls access to the asset by engaging or disengaging the system lock based on internal logic and information from the host computer.
- A host computer, which processes and stores the biometric trait in a database.
- Specialized software, which compares an individual image taken by the sensor with a stored profile or profiles.
- A locking mechanism, which is controlled by the biometric system.
- A power source to power the system.

Biometric Hand and Finger Geometry Recognition

Hand and finger geometry recognition is the process of identifying an individual through the unique geometry (shape, thickness, length, width, etc.) of that individual's hand or fingers. Hand geometry recognition has been employed since the early 1980s and is among the most widely used biometric technologies for controlling access to important assets. It is easy to install and use, and it is appropriate for use in any location requiring use of highly accurate, nonintrusion biometric security. For example, it is currently used in numerous workplaces, day care facilities, hospitals, universities, airports, and power plants.

A newer option within hand geometry recognition technology is finger geometry recognition (not to be confused with fingerprint recognition). Finger geometry recognition relies on the same scanning methods and technologies as does hand geometry recognition, but the scanner only scans two of the user's fingers, as opposed to the entire hand. Finger geometry recognition has been in commercial use since the mid-1990s and is mainly used in time and attendance applications (i.e., to track when individuals have entered and exited a location). To date the only large-scale commercial use of two-finger geometry for controlling access is at Disney World, where season pass holders use the geometry of their index and middle finger to gain access to the facilities.

To use a hand or finger geometry unit, an individual presents his or her hand or fingers to the biometric unit for scanning. The scanner consists of a charged coupled device (CCD), which is essentially a high-resolution digital camera; a reflective platen on which the hand is placed; and a mirror or mirrors that help capture different angles of the hand or fingers. The camera scans individual geometric characteristics of the hand or fingers by taking multiple images while the user's hand rests on the reflective platen. The camera also captures depth, or three-dimensional information, through light reflected from the mirrors and the reflective platen. This live image is then compared to a template that was previously established for that individual when he or she was enrolled in the system. If the live scan of the individual matches the stored template, the individual is "verified" and is given access to that asset. Typically, verification takes about two seconds. In access control applications, the scanner is usually connected to some sort of electronic lock, which unlocks the door, turnstile, or other entry barrier when the user is verified. The user can then proceed through the entrance. In time and attendance applications, the time that an individual checks in and out of a location is stored for later use.

As discussed above, hand and finger geometry recognition systems can be used in several different types of applications, including access control and time and attendance tracking. While time and attendance tracking can be used for security, it is primarily used for operations and payroll purposes (i.e., clocking in and clocking out). In contrast, access control applications are more likely to be security related. Biometric systems are widely used for access control and can be used on various types of assets, including entryways, computers, vehicles, etc. However, because of their size, hand/finger recognition systems are primarily used in entryway access control applications.

Biometric Overview—Iris Recognition

The iris, which is the colored or pigmented area of the eye surrounded by the sclera (the white portion of the eye), is a muscular membrane that controls the amount of light entering the eye by contracting or expanding the pupil (the dark center of the eye). The dense, unique patterns of connective tissue in the human iris were first noted in 1936, but it was not until 1994, when algorithms for iris recognition were created and patented, that commercial applications using biometric iris recognition began to be used extensively. There are now two vendors producing iris recognition technology—both the original developer of these algorithms and a second company, which has developed and patented a different set of algorithms for iris recognition.

The iris is an ideal characteristic for identifying individuals because it is formed *in utero*, and its unique patterns stabilize around eight months after birth. No two irises are alike; neither are an individual's right and left irises, nor the irises of identical twins. The iris is protected by the cornea (the clear covering over the eye), and therefore it is not subject to the aging or physical changes (and potential variation) that are common to some other biometric measures, such as the hand, fingerprints, and the face. Although some limited changes can occur naturally over time, these changes generally occur in the iris's melanin and therefore affect only the eye's color and not its unique patterns. (In addition, because iris scanning uses only black-and-white images, color changes would not affect the scan anyway.) Thus, barring specific injuries or certain rare surgeries directly affecting the iris, the iris's unique patterns remain relatively unchanged over an individual's lifetime.

Iris recognition systems employ a monochromatic or black-and-white video camera that uses both visible and near infrared light to take video of an individual's iris. Video is used rather than still photography as an extra security procedure. The video is used to confirm the normal continuous fluctuations of the pupil as the eye focuses, which ensures that the scan is of a living human being, and not a photograph or some other attempted hoax. A high-resolution image of the iris is then captured or extracted from the video, using a device often referred to as a "frame grabber." The unique characteristics identified in this image are then converted into a numeric code, which is stored as a template for that user.

Card Identification/Access/Tracking Systems

A card reader system is a type of electronic identification system that is used to identify a card and then perform an action associated with that card. Depending on the system, the card may identify where a person is or where he or she was at a certain time, or it may authorize another action, such as disengaging a lock. For example, a security guard may use his or her card at card readers located throughout a facility to indicate that he or she has checked a certain location at a certain time. The reader will store the information and/or send it to a central location, where it can be checked later to ensure that the guard has patrolled the area. Other card reader systems can be associated with a lock, so that the cardholder must have his or her card read and accepted by the reader before the lock disengages.

A complete card reader system typically consists of the following components:

- access cards that are carried by the user
- card readers, which read the card signals and send the information to control units

- control units, which control the response of the card reader to the card
- a power source

A "card" may be a typical card or another type of device, such as a key fob or wand. These cards store electronic information, which can range from a simple code (i.e., the alphanumeric code on a Proximity card) to individualized personal data (i.e., biometric data on a Smartcard). The card reader reads the information stored on the card and sends it to the control unit, which determines the appropriate action to take when a card is presented. For example, in a card access system, the control unit compares the information on the card to stored access authorization information to determine if the card holder is authorized to proceed through the door. If the information stored in the card reader system indicates that the key is authorized to allow entrance through the doorway, the system disengages the lock and the key holder can proceed through the door.

There are many different types of card reader systems on the market. The primary differences between card reader systems are in the way that data is encoded on the cards, in the way these data are transferred between the card and the card reader, and in the types of applications for which they are best suited. However, all card systems are similar in the way that the card reader and control unit interact to respond to the card.

While card readers are similar in the way that the card reader and control unit interact to control access, they are different in the way data is encoded on the cards and the way these data are transferred between the card and the card reader. There are several types of technologies available for card reader systems. These include the following:

- Proximity
- Wiegand
- Smartcard
- Magnetic Stripe
- Bar Code
- Infrared
- Barium Ferrite
- Hollerith
- Mixed Technologies

Table 9.8 summarizes various aspects of card reader technologies. The determination for the level of security rate (low, moderate, or high) is based on the level of technology a given card reader system has and how simple it is to duplicate that technology and thus bypass the security. Vulnerability ratings were based on whether the card reader can be damaged easily due to frequent use or difficult working conditions (i.e., weather conditions if the reader is located outside). Often this is influenced by the number of moving parts in the system—the more moving parts, then greater the system's potential susceptibility to damage. The life cycle rating is based on the durability of a given card reader system over its entire operational period. Systems requiring frequent physical contact between the reader and the card often have a shorter life cycle due to the wear and tear to which the equipment is exposed. For many card reader systems, the vulnerability rating and life cycle ratings have a reciprocal relationship. For instance, if a given system has a high vulnerability rating it will almost always have a shorter life cycle.

Card reader technology can be implemented for facilities of any size and with any number of users. However, because individual systems vary in the complexity of their technology and in the level of security they can provide to a facility, individual users must determine the appropriate system for their needs. Some important features to consider when selecting a card reader system include:

- The technological sophistication and security level of the card system.
- The size and security needs of the facility.
- The frequency with which the card system will be used. For systems that will experience a high frequency of use it is important to consider a system that has a longer life cycle and lower vulnerability rating, thus making it more cost effective to implement.
- The conditions in which the system will be used (i.e., will it be used on the interior or exterior of buildings, does it require light or humidity controls, etc.). Most card reader systems can operate under normal environmental conditions, and therefore this would be a mitigating factor only in extreme conditions.
- System costs.

Exterior Intrusion—Buried Sensors

Buried sensors are electronic devices that are designed to detect potential intruders. The sensors are buried along the perimeters of sensitive assets and are able to detect intruder activity both aboveground and belowground. Some of these systems are composed of individual, stand-alone sensor units, while other sensors consist of buried cables.

There are four types of buried sensors that rely on different types of triggers: pressure or seismic; magnetic field; ported coaxial cable; and fiber-optic cables. These four sensors are all covert and terrain-following, meaning they are hidden from view and follow the contour of the terrain. The four types of sensors are described in more detail below. Table 9.9 presents the distinctions between the four types of buried sensors.

Table 9.8. Card reader technology

Types of Security	*Technology*	*Life Cycle*	*Vulnerability*	*Level of Card Readers*
Proximity	Embedded radio frequency circuits encoded with unique information	Long	Virtually none	Moderate-high
Wiegand	Short lengths of small-diameter, special alloy wire with unique magnetic properties	Long	Low susceptibility to damage; high durability due to embedded wires	Moderate-expensive
Magnetic Stripe	Electromagnetic charges to encode information on a piece of tape attached to back of card	Moderate	Moderately susceptible to damage due to frequency of use	Low-moderate
Bar Code	Series of narrow and wide bars and spaces	Short	High; easily damaged	Low
Hollerith	Holes punched in a plastic or paper card and read optically	Short	High; easily damaged from frequent use	Low
Infrared	An encoded shadow pattern within the card, read using an infrared scanner	Moderate	IR scanners are optical and, thus, vulnerable to contamination.	High
Barium Ferrite	Uses small bits of magnetized barium ferrite, placed inside a plastic card. The polarity and location of the "spots" determine the coding.	Moderate	Low susceptibility to damage; durable since spots are embedded in the material	Moderate-high
SmartCard	Patterns or series of narrow and wide bars and spaces.	Short	High susceptibility to damage, low durability	Highest

Source: USEPA (2005).

Table 9.9. Types of buried sensors

Type	*Description*
Pressure or seismic	Responds to disturbances in the soil
Magnetic field	Responds to a change in the local magnetic field caused by the movement of nearby metallic material
Ported coaxial cables	Responds to motion of a material with a high dielectric constant or high conductivity near the cables
Fiber-optic cables	Responds to a change in the shape of the fiber that can be sensed using sophisticated sensors and computer signal processing

Source: Adapted from Garcia (2001).

Exterior Intrusion Sensors

An exterior intrusion sensor is a detection device that is used in an outdoor environment to detect intrusions into a protected area. These devices are designed to detect an intruder and then communicate an alarm signal to an alarm system. The alarm system can respond to the intrusion in many different ways, such as by triggering an audible or visual alarm signal, or by sending an electronic signal to a central monitoring location that notifies security personnel of the intrusion.

Intrusion sensors can be used to protect many kinds of assets. Intrusion sensors that protect physical space are classified according to whether they protect indoor (interior) space (i.e., an entire building or room within a building), or outdoor (exterior) space (i.e., a fence line or perimeter). Interior intrusion sensors are designed to protect the interior space of a facility by detecting an intruder who is attempting to enter, or who has already entered a room or building. In contrast, exterior intrusion sensors are designed to detect an intrusion into a protected outdoor/exterior area. Exterior protected areas are typically arranged as zones or exclusion areas placed so that the intruder is detected early in the intrusion attempt before the intruder can gain access to more valuable assets (e.g., into a building located within the protected area). Early detection creates additional time for security forces to respond to the alarm.

Exterior intrusion sensors are classified according to how the sensor detects the intrusion within the protected area. The three classes of exterior sensor technology include the following:

- buried-line sensors
- fence-associated sensors
- freestanding sensors

1. *Buried-line sensors.* As the name suggests, buried-line sensors are sensors that are buried underground and are designed to detect disturbances within the ground—

such as disturbances caused by an intruder digging, crawling, walking, or running on the monitored ground. Because they sense ground disturbances, these types of sensors are able to detect intruder activity both on the surface and below ground. Individual types of exterior buried-line sensors function in different ways, including by detecting motion, pressure, or vibrations within the protected ground; or by detecting changes in some type of field (e.g., a magnetic field) that the sensors generate within the protected ground. Specific types of buried-line sensors include pressure or seismic sensors, magnetic field sensors, ported coaxial cables, and fiber-optic cables. Details on each of these sensor types are provided below.

- *Buried-line pressure* or *seismic sensors* detect physical disturbances to the ground—such as vibrations or soil compression—caused by intruders walking, driving, digging, or otherwise physically contacting the protected ground. These sensors detect disturbances from all directions and, therefore, can protect an area radially outward from their location; however, because detection may weaken as a function of distance from the disturbance, choosing the correct burial depth from the design area will be crucial. In general, sensors buried at a shallow depth protect a relatively small area but have a high probability of detecting intrusion within that area, while sensors buried at a deeper depth protect a wider area but have a lower probability of detecting intrusion into that area.
- *Buried-line magnetic field sensors* detect changes in a local magnetic field that are caused by the movement of metallic objects within that field. This type of sensor can detect ferric metal objects worn or carried by an intruder entering a protected area on foot as well as vehicles being driven into the protected area.
- *Buried-line ported coaxial cable sensors* detect the motion of any object (i.e., human body, metal, etc.) possessing high conductivity and located within close proximity to the cables. An intruder entering into the protected space creates an active disturbance in the electric field, thereby triggering an alarm condition.
- *Buried-line fiber-optic cable sensors* detect changes in the attenuation of light signals transmitted within the cable. When the soil around the cable is compressed, the cable is distorted, and the light signal transmitted through the cable changes, initiating an alarm. This type of sensor is easy to install because it can be buried at a shallow burial depth (only a few centimeters) and still be effective.

2. *Fence-associated sensors.* Fence-associated sensors are either attached to an existing fence, or are installed in such a way as to create a fence. These sensors detect disturbances to the fence—such as those caused by an intruder attempting to climb the fence, or by an intruder attempting to cut or lift the fence fabric. Exterior fence-associated sensors include fence-disturbance sensors, taut-wire sensor fences, and electric field or capacitance sensors. Details on each of these sensor types are provided below.

- *Fence-disturbance sensors* detect the motion or vibration of a fence, such as that caused by an intruder attempting to climb or cut through the fence. In general, fence-disturbance sensors are used on chain-link fences or on other fence types where a movable fence fabric is hung between fence posts.
- *Taut-wire sensor fences* are similar to fence-disturbance sensors except that instead of attaching the sensors to a loose fence fabric, the sensors are attached to a wire that is stretched tightly across the fence. These types of systems are designed to detect changes in the tension of the wire rather than vibrations in the fence fabric. Taut-wire sensor fences can be installed over existing fences, or as standalone fence systems.
- *Electric field* or *capacitance sensors* detect changes in capacitive coupling between wires that are attached to, but electrically isolated from, the fence. As opposed to other fence-associated intrusion sensors, both electric field and capacitance sensors generate an electric field that radiates out from the fence line, resulting in an expanded zone of protection relative to other fence-associated sensors, and allowing the sensor to detect the presence of intruders before they arrive at the fence line. (Note: Proper spacing is necessary during installation of the electric field sensor to prevent a would-be intruder from slipping between widely spaced wires.)

3. *Free-standing sensors*—These sensors, which include active infrared, passive infrared, bistatic microwave, monostatic microwave, dual-technology, and video motion detection (VMD) sensors, consist of individual sensor units or components that can be set up in a variety of configurations to meet a user's needs. They are installed aboveground, and depending on how they are oriented relative to each other, they can be used to establish a protected perimeter or a protected space. More details on each of these sensor types are provided below.
   - *Active infrared sensors* transmit infrared energy into the protected space and monitor for changes in this energy caused by intruders entering that space. In a typical application, an infrared light beam is transmitted from a transmitter unit to a receiver unit. If an intruder crosses the beam, the beam is blocked, and the receiver unit detects a change in the amount of light received, triggering an alarm. Different sensors can see single- and multiple-beam arrays. Single-beam infrared sensors transmit a single infrared beam. In contrast, multiple-beam infrared sensors transmit two or more beams parallel to each other. This multiple-beam sensor arrangement creates an infrared "fence."
   - *Passive infrared (PIR) sensors* monitor the ambient infrared energy in a protected area, and evaluate changes in that ambient energy that may be caused by intruders moving through the protected area. Detection ranges can exceed 100 yards on cold days, with size and distance limitations dependent upon the background temperature. PIR sensors generate a non-uniform detection pattern (or "cur-

tain") that has areas (or "zones") of more sensitivity and areas of less sensitivity. The specific shape of the protected area is determined by the detector's lenses. The general shape common to many detection patterns is a series of long "fingers" emanating from the PIR and spreading in various directions. When intruders enter the detection area, the PIR sensor detects differences in temperature due to the intruders' body heat, and triggers an alarm. While the PIR leaves unprotected areas between its fingers, an intruder would be detected if he passed from a non-protected area to a protected area.

- *Microwave sensors* detect changes in received energy generated by the motion of an intruder entering into a protected area. Monostatic microwave sensors incorporate a transmitter and a receiver in one unit, while bistatic sensors separate the transmitter and the receiver into different units. Monostatic sensors are limited to a coverage area of 400 feet, while bistatic sensors can cover an area up to 1,500 feet. For bistatic sensors, a zone of no detection exists in the first few feet in front of the antennas. This distance from the antennas to the point at which the intruder is first detected is known as the offset distance. Due to this offset distance, antennas must be configured so that they overlap one another (as opposed to being adjacent to each other), thereby creating long perimeters with a continuous line of detection.
- *Dual-technology sensors* consist of two different sensor technologies incorporated together into one sensor unit. For example, a dual technology sensor could consist of a passive infrared detector and a monostatic microwave sensor integrated into the same sensor unit.
- *Video motion detection* (VMD) *sensors* monitor video images from a protected area for changes in the images. Video cameras are used to detect unauthorized intrusion into the protected area by comparing the most recent image against a previously established one. Cameras can be installed on towers or other tall structures so that they can monitor a large area.

Fences

A fence is a physical barrier that can be set up around the perimeter of an asset. Fences often consist of individual pieces (such as individual pickets in a wooden fence, or individual sections of a wrought-iron fence) that are fastened together. Individual sections of the fence are fastened together using posts, which are sunk into the ground to provide stability and strength for the sections of the fence hung between them. Gates are installed between individual sections of the fence to allow access inside the fenced area.

Many fences are used as decorative architectural features to separate physical spaces from each other. They may also be used to physically mark the location of a boundary (such as a fence installed along a property line). However, a fence can also serve as an effective means for physically delaying intruders from gaining access to an industrial chemical facility. For example, many utilities install fences around their primary facilities, around remote pump stations, or around hazardous materials storage areas or sensitive areas within a facility. Access to the area can be controlled through security at gates or doors through the fence (for example, by posting a guard at the gate or by locking it). In order to gain access to the asset, unauthorized persons would either have to go around or through the fence.

Fences are often compared with walls when determining the appropriate system for perimeter security. While both fences and walls can provide adequate perimeter security, fences are often easier and less expensive to install than walls. However, they do not usually provide the same physical strength that walls do. In addition, many types of fences have gaps between the individual pieces that make up the fence (i.e., the spaces between chain links in a chain-link fence or the space between pickets in a picket fence). Thus, many types of fences allow the interior of the fenced area to be seen. This may allow intruders to gather important information about the locations or defenses of vulnerable areas within the facility.

There are numerous types of materials used to construct fences, including chain-link iron, aluminum, wood, or wire. Some types of fences, such as split rails or pickets, may not be appropriate for security purposes because they are traditionally low fences, and they are not physically strong. Potential intruders may be able to easily defeat these fences either by jumping or climbing over them or by breaking through them. For example, the rails in a split fence could be broken easily.

Important security attributes of a fence include the height to which it can be constructed, the strength of the material comprising the fence, the method and strength of attaching the individual sections of the fence together at the posts, and the fence's ability to restrict the view of the assets inside the fence. Additional considerations should include the ease of installing the fence and the ease of removing and reusing sections of the fence. Table 9.10 provides a comparison of the important security and usability features of various fence types.

Some fences can include additional measures to delay, or even detect, potential intruders. Such measures may include the addition of barbed wire, razor wire, or other

Table 9.10. Comparison of different fence types

Specifications	*Chain Link*	*Iron*	*Wire (Wirewall)*	*Wood*
Height limitations	12′	12′	12′	8′
Strength	Medium	High	High	Low
Installation requirements	Low	High	High	Low
Ability to remove/reuse	Low	High	Low	High
Ability to replace/repair	Medium	High	Low	High

Source: USEPA (2005).

deterrents at the top of the fence. Barbed wire is sometimes employed at the base of fences as well. This can impede a would-be intruder's progress in even reaching the fence. Fences may also be fitted with security cameras to provide visual surveillance of the perimeter. Finally, some facilities have installed motion sensors along their fences to detect movement on the fence. Several manufacturers have combined these multiple perimeter security features into one product and offer alarms and other security features.

The correct implementation of a fence can make it a much more effective security measure. Security experts recommend the following when a facility constructs a fence:

- The fence should be at least seven to nine feet high.
- Any outriggers, such as barbed wire, that are affixed on top of the fence should be angled out and away from the facility, and not in toward the facility. This will make climbing the fence more difficult, and will prevent ladders from being placed against the fence.
- Other types of hardware can increase the security of the fence. This can include installing concertina wire along the fence (this can be done in front of the fence or at the top of the fence), or adding intrusion sensors, cameras, or other hardware to the fence.
- All undergrowth should be cleared for several feet (typically six feet) on both sides of the fence. This will allow for a clearer view of the fence by any patrols in the area.
- Any trees with limbs or branches hanging over the fence should be trimmed so that intruders cannot use them to go over the fence. Also, it should be noted that fallen trees can damage fences, and so management of trees around the fence can be important. This can be especially important where fence goes through a remote area.
- Fences that do not block the view from outside the fence to inside the fence allow patrols to see inside the fence without having to enter the facility.
- "No Trespassing" signs posted along the fence can be a valuable tool in prosecuting any intruders who claim that the fence was broken and that they did not enter through the fence illegally. Adding signs that highlight the local ordinances against trespassing can further dissuade simple troublemakers from illegally jumping/climbing the fence.

Films for Glass Shatter Protection

Most industrial chemical complexes have numerous windows on the outside of buildings, in doors, and in interior offices. In addition, many facilities have glass doors or other glass structures, such as glass walls or display cases. These glass objects are potentially vulnerable to shattering when heavy objects are thrown or launched at them, when explosions occur near them, or when there are high winds (for exterior glass). If the glass is shattered, intruders may potentially enter an area. In addition, shattered glass projected into a room from an explosion or from an object being thrown through a door or window can injure and potentially incapacitate personnel in the room. Materials that prevent glass from shattering can help to maintain the integrity of the door, window, or other glass object, and can delay an intruder from gaining access. These materials can also prevent flying glass and thus reduce potential injuries.

Materials designed to prevent glass from shattering include specialized films and coatings. These materials can be applied to existing glass objects to improve their strength and their ability to resist shattering. The films have been tested against many scenarios that could result in glass breakage, including penetration by blunt objects, bullets, high winds, and simulated explosions. Thus, the films are tested against both simulated weather scenarios (which could include both the high winds themselves and the force of objects blown into the glass), as well as more criminal/terrorist scenarios where the glass is subject to explosives or bullets. Many vendors provide information on the results of these types of tests, and thus potential users can compare different product lines to determine which products best suit their needs.

The primary attributes of films for shatter protection are the following:

- the materials from which the film is made
- the adhesive that bonds the film to the glass surface
- the thickness of the film

Standard glass safety films are designed from high-strength polyester. Polyester provides both strength and elasticity, which is important in absorbing the impact of an object, spreading the force of the impact over the entire film, and resisting tearing. The polyester is also designed to be resistant to scratching, which can result when films are cleaned with abrasives or other industrial cleaners.

The bonding adhesive is important in ensuring that the film does not tear away from the glass surface. This can be especially important when the glass is broken, so that the film does not peel off the glass and allow it to shatter. In addition, films applied to exterior windows can be subject to high concentrations of UV light, which can break down bonding materials.

Film thickness is measured in gauge or mils. According to test results reported by several manufacturers, film thickness appears to affect resistance to penetration/tearing, with thicker films being more resistant to penetration and tearing. However, the application of a thicker film did not decrease glass fragmentation.

Many manufacturers offer films in different thicknesses. The standard film is usually one four-mil layer; thicker films are typically composed of several layers of the standard four-mil sheet. However, newer technologies have allowed the polyester to be "microlayered" to produce a stronger film without significantly increasing its thickness. In this microlayering process, each laminate film is composed of multiple micro-thin layers of polyester woven together at alternating angles. This provides increased strength for the film, while maintaining the flexibility and thin profile of one film layer.

As described above, many vendors test their products in various scenarios that would lead to glass shattering, including simulated bomb blasts and simulation of the glass being struck by wind-blown debris. Some manufacturers refer to the Government Services Administration standard for bomb blasts, which requires resistance to tearing for a four-psi blast. Other manufacturers use other measures and tests for resistance to tearing. Many of these tests are not "standard," in that no standard testing or reporting methods have been adopted by any of the accepted standards-setting institutions. However, many of the vendors publish the procedure and the results of these tests on their websites, and this may allow users to evaluate the protectiveness of these films. For example, several vendors evaluate the "protectiveness" of their films and the "hazard" resulting from blasts near windows with and without protective films. Protectiveness is usually evaluated based on the percentage of glass ejected from the window, and the height at which that ejected glass travels during the blast (for example, if the blasted glass tends to project upward into a room—potentially toward people's faces—it is a higher hazard than if it is blown downward into the room toward people's feet). There are some standard measures of glass breakage. For example, several vendors indicated that their products exceed American Society for Testing and Materials (ASTM) standard 64Z-95 "Standard Test Method for Glazing and Glazing Systems Subject to Air Blast Loadings." Vendors often compare the results of some sort of penetration or force tests, ballistic tests, or simulated explosions with unprotected glass versus glass onto which their films have been applied. Results generally show that applying films to the glass surfaces reduces breakage/penetration of the glass and can reduce the amount and direction of glass ejected from the frame. This in turn reduces the hazard from flying glass.

In addition to these types of tests, many vendors conduct standard physical tests on their products, such as tests for tensile strength and peel strength. Tensile strength indicates the strength per area of material, while the peel strength indicates the force it would take to peel the product from the glass surface. Several vendors indicate that their products exceed American National Standards Institute (ANSI) standard Z97.1 for tensile strength and adhesion.

Vendors typically have a warranty against peeling or other forms of deterioration of their products. However, the warranty requires that the films be installed by manufacturer-certified technicians to ensure that they are applied correctly and therefore that the warranty is in effect. Warranties from different manufacturers may vary. Some may cover the cost of replacing the material only, while others include material plus installation. Because installation costs are significantly greater than material costs, different warranties may represent large differences in potential costs.

Fire Hydrant Locks

Fire hydrants are installed at strategic locations throughout a community's water distribution system to supply water for firefighting. However, because there are many hydrants in a system and they are often located in residential neighborhoods, industrial districts, and other areas where they cannot be easily observed and/or guarded, they are potentially vulnerable to unauthorized access. Many municipalities, states, and EPA regions have recognized this potential vulnerability and have instituted programs to lock hydrants. For example, EPA Region 1 has included locking hydrants as number seven on its "Drinking Water Security and Emergency Preparedness" top ten list for small groundwater suppliers.

A hydrant lock is a physical security device designed to prevent unauthorized access to the water supply through a hydrant. It can also ensure water and water pressure availability to firefighters and prevent water theft and associated lost water revenue. These locks have been successfully used in numerous municipalities and in various climates and weather conditions.

Fire hydrant locks are basically steel covers or caps that are locked in place over the operating nut of a fire hydrant. The lock prevents unauthorized persons from accessing the operating nut and opening the fire hydrant valve. The lock also makes it more difficult to remove the bolts from the hydrant and access the system that way. Finally, hydrant locks shield the valve from being broken off. Should a vandal attempt to breach the hydrant lock by force and succeed in breaking the hydrant lock, the vandal will only succeed in bending the operating valve. If the hydrant's operating valve is bent, the hydrant will not be operational, but the water asset remains protected and inaccessible to vandals. However, the entire hydrant will need to be replaced.

Hydrant locks are designed so that the hydrants can be operated by special key wrenches without removing the lock. These specialized wrenches are generally distributed to the fire department, public works department, and other authorized persons so that they can access the hydrants as needed. An inventory of wrenches and their serial numbers is generally kept by a municipality so that the location of all wrenches is known. These operating key wrenches may only be purchased by registered lock owners.

The most important features of hydrant locks are their strength and the security of their locking systems. The locks must be strong so that they cannot be broken off. Hydrant locks are constructed from stainless or alloyed steel. Stainless steel locks are stronger and are ideal for all climates; however, they are more expensive than alloy locks. The locking mechanisms for each fire hydrant locking system ensure that the hydrant can only be operated by authorized personnel who have the specialized key to work the hydrant.

Hatch Security

A hatch is basically a door that is installed on a horizontal plane (such as in a floor, a paved lot, or a ceiling), instead of on a vertical plane (such as in a building wall). Hatches are usually used to provide access to assets that are either located underground (such as hatches to basements or underground storage areas) or to assets located above ceilings (such as emergency roof exits). At chemical industrial facilities, hatches are typically used to provide access to underground vaults containing pumps, meter chambers, valves, or piping, or to the interior of chemical tanks or covered water reservoirs. Securing a hatch by locking it or upgrading materials to give the hatch added strength can help to delay unauthorized access to any asset behind the hatch.

Like all doors, a hatch consists of a frame anchored to the horizontal structure, a door or doors, hinges connecting the door(s) to the frame, and a latching or locking mechanism that keeps the hatch door(s) closed. It should be noted that improving hatch security is straightforward, and that hatches with upgraded security features can be installed new or they can be retrofit for existing applications. Depending on the application, the primary security-related attributes of a hatch are the strength of the door and frame, its resistance to the elements and corrosion, its ability to be sealed against water or gas, and its locking features.

Hatches must be both strong and lightweight so that they can withstand typical static loads (such as people or vehicles walking or driving over them) while still being easy to open. In addition, because hatches are typically installed at outdoor locations, they are usually designed from corrosion-resistant metal that can withstand the elements. Therefore, hatches are typically constructed from high-gauge steel or lightweight aluminum.

Aluminum is typically the material of choice for hatches because it is lightweight and more corrosion-resistant than steel. Aluminum is not as rigid as steel, so aluminum hatch doors may be reinforced with aluminum stiffeners to provide extra strength and rigidity. The doors are usually constructed from single or double layers (or "leaves") of material. Single-leaf designs are standard for smaller hatches, while double-leaf designs are required for larger hatches. In addition, aluminum products do not require painting. This is reflected in the warranties available with different products. Product warranties range from ten years to lifetime.

Steel is heavier per square foot than aluminum, and thus steel hatches will be heavier and more difficult to open than aluminum hatches of the same size. However, heavy steel hatch doors may have spring-loaded, hydraulic, or gas openers or other specialized features that help in opening the hatch and in keeping it open.

Many hatches are installed in outdoor areas, often in roadways or pedestrian areas. Therefore, the hatch installed for any given application must be designed to withstand the expected load at that location. Hatches are typically sold to withstand either pedestrian or vehicle loading. Pedestrian-loading hatches are typically designed to withstand either 150 or 300 pounds per square foot (psf) of loading. The vehicle-loading standard is the American Association of State Highway and Transportation Officials (AASHTO) H-20 wheel-loading standard of 16,000 lbs over an eight-inch by twenty-inch area. It should be noted that these design parameters are for static loads and not dynamic loads; thus, the loading capabilities may not reflect potential resistance to other types of loads that may be more typical of an intentional threat, such as repeated blows from a sledgehammer or pressure generated by bomb blasts or bullets.

The typical design for a watertight hatch includes a channel frame that directs water away from the hatch. This can be especially important in a hatch on a storage tank, because this will prevent liquid contaminants from being dumped on the hatch and leaking through into the interior. Hatches can also be constructed with gasket seals that are air-, odor-, and gas-tight.

Typically, hatches for pedestrian-loading applications have hinges located on the exterior of the hatch, while hatches designed for H-20 loads have hinges located in the interior of the hatch. Hinges located on the exterior of the hatch could be removed, thereby allowing intruders to remove the hatch door and access the asset behind the hatch. Therefore, installing H-20 hatches even for applications that do not require H-20 loading levels may increase security, because intruders will not be able to tamper with the hinges and circumvent the hatch this way.

In addition to the location of the hinges, stock hinges can be replaced with heavy-duty or security hinges that are more resistant to tampering.

The hatch locking mechanism is perhaps the most important part of hatch security. There are a number of locks that can be implemented for hatches, including the following:

- slam locks (internal locks that are located within the hatch frame)
- recessed cylinder locks
- bolt locks
- padlocks

Ladder Access Control

Chemical industrial facilities have a number of assets that are raised above ground level, including raised water tanks, raised chemical tanks, raised piping systems, and roof access points into buildings. In addition, communications equipment, antennae, or other electronic devices may be located on the top of these raised assets. Typically, these assets are reached by ladders that are permanently anchored to the asset. For example, raised chemical/water tanks typically are accessed by ladders that are bolted to one of the legs of the tank. Controlling access to these raised assets by controlling access to the ladder can increase security at a chemical industrial facility.

A typical ladder access control system consists of some type of cover that is locked or secured over the ladder. The cover can be a casing that surrounds most of the ladder, or a door or shield that covers only part of the ladder. In either case, several rungs of the ladder (the number of rungs depends on the size of the cover) are made inaccessible by the cover, and these rungs can only be accessed by opening or removing the cover. The cover is locked so that only authorized personnel can open or remove it and use the ladder. Ladder access controls are usually installed at several feet above ground level, and they usually extend several feet up the ladder so that they cannot be circumvented by someone accessing the ladder above the control system.

The important features of ladder access control are the size and strength of the cover and its ability to lock or otherwise be secured from unauthorized access.

The covers are constructed from aluminum or some type of steel. This should provide adequate protection from being pierced or cut through. The metals are corrosion resistant so that they will not corrode or become fragile from extreme weather conditions in outdoor applications. The bolts used to install each of these systems are galvanized steel. In addition, the bolts for each cover are installed on the inside of the unit so they cannot be removed from the outside.

Locks

A lock is a type of physical security device that can be used to delay or prevent a door, a window, a manhole, a filing cabinet drawer, or some other physical feature from being opened, moved, or operated. Locks typically operate by connecting two pieces together—such as by connecting a door to a door jamb or a manhole to its casement. Every lock has two modes—engaged (locked) and disengaged (opened). When a lock is disengaged, the asset on which the lock is installed can be accessed by anyone, but when the lock is engaged, only authorized personnel can access the locked asset.

Locks are excellent security features because they have been designed to function in many ways and to work on many different types of assets. Locks can also provide different levels of security depending on how they are designed and implemented. The security provided by a lock is dependent on several factors, including its ability to withstand physical damage (e.g., whether it can be cut off, broken, or otherwise physically disabled) as well as its requirements for supervision or operation (e.g., combinations may need to be changed frequently so that they are not compromised and the locks remain secure). While there is no single definition of the "security" of a lock, locks are often described as minimum, medium, or maximum security. Minimum-security locks are those that can be easily disengaged (or "picked") without the correct key or code, or those that can be disabled easily (such as small padlocks that can be cut with bolt cutters). Higher-security locks are more complex and thus are more difficult to pick, or are sturdier and more resistant to physical damage.

Many locks, such as many door locks, only need to be unlocked from one side. For example, most door locks need a key to be unlocked only from the outside. A person opens such devices, called single-cylinder locks, from the inside by pushing a button or by turning a knob or handle. Double-cylinder locks require a key to be locked or unlocked from both sides.

Manhole Intrusion Sensors

Manholes are commonly found in chemical industrial sites. Manholes are designed to provide access to underground utilities, meter vaults, chemical pumping rooms, and so forth, and therefore they are potential entry points to a system. Because many utilities run under other infrastructure (roads, buildings), manholes also provide potential access points to critical infrastructure as well as chemical process assets. In addition, because the portion of the system to which manholes provide entry is primarily located underground, access to a system through a manhole increases the chance that an intruder will not be seen. Therefore, protecting manholes can be a critical component of guarding an entire plant site and a surrounding community.

There are multiple methods for protecting manholes, including preventing unauthorized personnel from physically accessing the manhole and detecting attempts at unauthorized access to the manhole.

A manhole intrusion sensor is a physical security device designed to detect unauthorized access to the facility through a manhole. Monitoring a manhole that provides access to a chemical plant or processing system can mitigate two distinct types of threats. First, monitoring a manhole may detect access of unauthorized personnel to chemical systems or assets through the manhole. Second, monitoring manholes may also allow the detection of intruders attempting to place explosive or other destructive (WMD) devices into the chemical system.

Several different technologies have been used to develop manhole intrusion sensors, including mechanical systems, magnetic systems, and fiber optic and infrared sensors. Some of these intrusion sensors have been specifically designed for manholes, while others consist of standard, off-the-shelf intrusion sensors that have been implemented in a system specifically designed for application in a manhole.

Manhole Locks

A manhole lock is a physical security device designed to delay unauthorized access to the chemical facility or system through a manhole.

Radiation Detection Equipment for Monitoring Personnel and Packages

One of the major potential threats facing any critical production facility or system is contamination by radioactive substances. Radioactive substances brought on-site at a facility could be used to contaminate the facility, thereby preventing workers from safely entering the facility. In addition, radioactive substances brought on-site at a chemical processing plant could be discharged into the chemical waste system, contaminating the downstream water supply. Therefore, detection of radioactive substances as they are brought on-site can be an important security enhancement.

Different radionuclides have unique properties, and different equipment is required to detect different types of radiation. However, it is impractical and potentially unnecessary to monitor for specific radionuclides carried on-site. Instead, for security purposes, it may be more useful to monitor for gross radiation as an indicator of unsafe substances.

In order to protect against these radioactive materials being brought on-site, a facility may set up monitoring sites outfitted with radiation detection instrumentation at entrances to the facility. Depending on the specific types of equipment chosen, this equipment would detect radiation emitted from people, packages, or other objects transported through an entrance.

One of the primary differences between the different types of detection equipment is the means by which the equipment reads the radiation. Radiation may be detected either by direct measurement or through sampling.

Direct radiation measurement involves measuring radiation through an external probe on the detection instrumentation. Some direct measurement equipment detects radiation emitted into the air around the monitored object. Because this equipment detects radiation in the air, it does not require that the monitoring equipment make physical contact with the monitored object. Direct means for detecting radiation include using a walk-through portal-type monitor that detects elevated radiation levels on a person or in a package, or by using a handheld detector, which is moved or swept over individual objects to locate a radioactive source.

Some types of radiation, such as alpha or low-energy beta radiation, have a short range and are easily shielded by various materials. These types of radiation cannot be measured through direct measurement. Instead, they must be measured through sampling. Sampling involves wiping the surface to be tested with a special filter cloth, and then reading the cloth in a special counter. For example, specialized smear counters measure alpha and low-energy beta radiation.

Reservoir Covers

Reservoirs are used to store raw or untreated water for chemical processing. They can be located underground (buried), at ground level, or on an elevated surface. Reservoirs can vary significantly in size; small reservoirs can hold as little as a thousand gallons, while larger reservoirs may hold many million gallons.

Reservoirs can be either natural or man-made. Natural reservoirs can include lakes or other contained water bodies, while man-made reservoirs usually consist of some sort of engineered structure, such as a tank or other impoundment structure. In addition to the water containment structure itself, reservoir systems may also include associated water treatment and distribution equipment, including intakes, pumps, pump houses, piping systems, chemical treatment and chemical storage areas, and so forth.

One of the most serious potential threats to the system is direct contamination of the stored water through dumping contaminants into the reservoir. Chemical facilities have taken various measures to mitigate this type of threat, including fencing off the reservoir, installing cameras to monitor for intruders, and monitoring for changes in water quality. Another option for enhancing security is covering the reservoir using some type of manufactured cover to prevent intruders from gaining physical access to the stored water. Implementing a reservoir cover may or may not be practical depending on the size of the reservoir (for example, covers are not typically used on natural reservoirs because they are too large for the cover to be technically feasible and cost effective).

A reservoir cover is a structure installed on or over the surface of the reservoir to minimize water-quality degradation. The three basic design types for reservoir covers are the following:

- floating
- fixed
- air-supported

A variety of materials are used when manufacturing a cover, including reinforced concrete, steel, aluminum, polypropylene, chlorosulfonated polyethylene, or ethylene interpolymer alloys. There are several factors that affect a reservoir cover's effectiveness and thus its ability to protect the stored water. These factors include the following:

- the location, size, and shape of the reservoir
- the ability to lay/support a foundation (for example, footing, soil, and geotechnical support conditions)
- the length of time the reservoir can be removed from service for cover installation or maintenance

- aesthetic considerations
- economic factors, such as capital and maintenance costs

For example, it may not be practical to install a fixed cover over a reservoir if the reservoir is too large or if the local soil conditions cannot support a foundation. A floating or air-supported cover may be more appropriate for these types of applications.

In addition to the practical considerations for installation of these types of covers, there are a number of operations and maintenance (O&M) concerns that affect the utility of a cover for specific applications, including how different cover materials will withstand local climatic conditions, what types of cleaning and maintenance will be required for each particular type of cover, and how these factors will affect the cover's lifespan and its ability to be repaired when it is damaged.

The primary feature affecting the security of a reservoir cover is its ability to maintain its integrity. Any type of cover, no matter what its construction material, will provide good protection from contamination by rainwater or atmospheric deposition, as well as from intruders attempting to access the stored water with the intent of causing intentional contamination. The covers are large and heavy, and it is difficult to circumvent them to get into the reservoir. At the very least, it would take a determined intruder, as opposed to a vandal, to defeat the cover.

Passive Security Barriers

One of the most basic threats facing any facility is from intruders accessing the facility with the intention of causing damage to its assets. These threats may include intruders actually entering the facility, as well as intruders attacking the facility from outside without actually entering it (i.e., detonating a bomb near enough to the facility to cause damage within its boundaries).

Security barriers are one of the most effective ways to counter the threat of intruders accessing a facility or the facility perimeter. Security barriers are large, heavy structures that are used to control access through a perimeter by either vehicles or personnel. They can be used in many different ways depending on how/where they are located at the facility. For example, security barriers can be used on or along driveways or roads to direct traffic to a checkpoint (e.g., a facility may install jersey barriers in a road to direct traffic in a certain direction). Other types of security barriers (crash beams, gates) can be installed at the checkpoint so that guards can regulate which vehicles can access the facility. Finally, other security barriers (e.g., bollards or security planters) can be used along the facility perimeter to establish a protective buffer area between the facility and approaching vehicles. Establishing such a protective buffer can help in mitigating the effects of the type of bomb blast described above, both by potentially absorbing some of the blast, and also by increasing the "stand-off" distance between the blast and the facility. (The force of an explosion is reduced as the shock wave travels farther from the source, and thus the farther the explosion is from the target, the less effective it will be in damaging the target.)

Security barriers can be either "active" or "passive." "Active" barriers, which include gates, retractable bollards, wedge barriers, and crash barriers, are readily movable, and thus they are typically used in areas where they must be moved often to allow vehicles to pass—such as in roadways at entrances and exits to a facility. In contrast to active security barriers, "passive" security barriers, which include jersey barriers, bollards, and security planters, are not designed to be moved on a regular basis, and thus they are typically used in areas where access is not required or allowed—such as along building perimeters or in traffic control areas. Passive security barriers are typically large, heavy structures that are usually several feet high, and they are designed so that even heavy-duty vehicles cannot go over or through them. Therefore, they can be placed in a roadway parallel to the flow of traffic so that they direct traffic in a certain direction (such as to a guardhouse, a gate, or some other sort of checkpoint), or perpendicular to traffic such that they prevent a vehicle from using a road or approaching a building or area.

Security for Doorways—Side-Hinged Doors

Doorways are the main access points to a facility or to rooms within a building. They are used on the exterior or in the interior of buildings to provide privacy and security for the areas behind them. Different types of doorway security systems may be installed in different doorways depending on the needs or requirements of the buildings or rooms. For example, exterior doorways tend to have heavier doors to withstand the elements and to provide some security to the entrance of the building. Interior doorways in office areas may have lighter doors that may be primarily designed to provide privacy rather than security. Therefore, these doors may be made of glass or lightweight wood. Doorways in industrial areas may have sturdier doors than do other interior doorways and may be designed to provide protection or security for areas behind the doorway. For example, fireproof doors may be installed in chemical storage areas or in other areas where there is a danger of fire.

Because they are the main entries into a facility or a room, doorways are often prime targets for unauthorized entry into a facility or an asset. Therefore, securing doorways may be a major step in providing security at a facility.

A doorway includes four main components:

- The door, which blocks the entrance. The primary threat to the actual door is breaking or piercing through the door. Therefore, the primary security features of doors are their strength and resistance to various physical threats, such as fire or explosions.

- The door frame, which connects the door to the wall. The primary threat to a door frame is that the door can be pried away from the frame. Therefore, the primary security feature of a door frame is its resistance to prying.
- The hinges, which connect the door to the door frame. The primary threat to door hinges is that they can be removed or broken, which will allow intruders to remove the entire door. Therefore, security hinges are designed to be resistant to breaking. They may also be designed to minimize the threat of removal from the door.
- The lock, which connects the door to the door frame. Use of the lock is controlled through various security features, such as keys, combinations, and so forth, such that only authorized personnel can open the lock and go through the door. Locks may also incorporate other security features, such as software or other systems, to track overall use of the door or to track individuals using the door, and so on.

Each of these components is integral in providing security for a doorway, and upgrading the security of only one of these components while leaving the other components unprotected may not increase the overall security of the doorway. For example, many facilities upgrade door locks as a basic step in increasing the security of a facility. However, if the facilities do not also focus on increasing security for the door hinges or the door frame, the door may remain vulnerable to being removed from its frame, thereby defeating the increased security of the door lock.

The primary attribute for the security of a door is its strength. Many security doors are 4-20 gauge hollow metal doors consisting of steel plates over a hollow cavity reinforced with steel stiffeners to give the door extra stiffness and rigidity. This increases resistance to blunt force used to try to penetrate through the door. The space between the stiffeners may be filled with specialized materials to provide fire, blast, or bullet resistance to the door.

The Windows and Doors Manufacturers Association has developed a series of performance attributes for doors. These include

- structural resistance
- forced-entry resistance
- hinge style screw resistance
- split resistance
- hinge resistance
- security rating
- fire resistance
- bullet resistance
- blast resistance

The first five bullet points provide information on a door's resistance to standard physical breaking and prying attacks. These tests are used to evaluate the strength of the door and the resistance of the hinges and the frame in a standardized way. For example, the rack load test simulates a prying attack on a corner of the door. A test panel is restrained at one end, and a third corner is supported. Loads are applied and measured at the fourth corner. The door impact test simulates a battering attack on a door and frame using impacts of two hundred foot-pounds by a steel pendulum. The door must remain fully operable after the test. It should be noted that door glazing is also rated for resistance to shattering, and so on. Manufacturers will be able to provide security ratings for these features of a door as well.

Door frames are an integral part of doorway security because they anchor the door to the wall. Door frames are typically constructed from wood or steel, and they are installed such that they extend for several inches over the doorway that has been cut into the wall. For added security, frames can be designed to have varying degrees of overlap with, or wrapping over, the underlying wall. This can make prying the frame from the wall more difficult. A frame formed from a continuous piece of metal (as opposed to a frame constructed from individual metal pieces) will prevent prying between pieces of the frame.

Many security doors can be retrofit into existing frames; however, many security door installations include replacing the door frame as well as the door itself. For example, bullet resistance per Underwriters Laboratories (UL) 752 requires resistance of the door and frame assembly, and thus replacing the door only would not meet UL 752 requirements.

Valve Lockout Devices

Valves are utilized as control elements in chemical process piping networks. They regulate the flow of both liquids and gases by opening, closing, or obstructing a flow passageway. Valves are typically located where flow control is necessary. They can be located in-line or at pipeline and tank entrance and exit points. They can serve multiple purposes in a process pipe network, including the following:

- redirecting and throttling flow
- preventing backflow
- shutting off flow to a pipeline or tank (for isolation purposes)
- releasing pressure
- draining extraneous liquid from pipelines or tanks
- introducing chemicals into the process network
- as access points for sampling process water

Valves may be located either aboveground or belowground. It is critical to provide protection against valve tampering. For example, tampering with a pressure relief valve could result in a pressure buildup and potential explosion in the piping network. On a larger scale, addition of a contaminant or non-compatible chemical substance to the chemical processing system through an unprotected valve could result in the catastrophic release of that contaminant to the general population.

Different security products are available to protect aboveground versus belowground valves. For example, valve lockout devices can be purchased to protect valves and valve controls located aboveground. Vaults containing underground valves can be locked to prevent access to these valves.

As described above, a lockout device can be used as a security measure to prevent unauthorized access to aboveground valves located within chemical processing systems. Valve lockout devices are locks that are specially designed to fit over valves and valve handles to control their ability to be turned or seated. These devices can be used to lock the valve into the desired position. Once the valve is locked, it cannot be turned unless the locking device is unlocked or removed by an authorized individual.

Various valve lockout options are available for industrial use, including:

- cable lockouts
- padlocked chains/cables
- valve-specific lockouts

Many of these lockout devices are not specifically designed for use in the chemical industry (e.g., chains, padlocks) and are available from a local hardware store or manufacturer specializing in safety equipment. Other lockout devices (for example, valve-specific lockouts or valve box-locks) are more specialized and must be purchased from safety or valve-related equipment vendors.

The three most common types of valves for which lockout devices are available are gate, ball, and butterfly valves. Each is described in more detail below.

- *Gate valve lockouts.* Gate valve lockouts are designed to fit over the operating hand wheel of the gate valve to prevent it from being turned. The lockout is secured in place with a padlock. Two types of gate valve lockouts are available: diameter-specific and adjustable. Diameter-specific lockouts are available for handles ranging from one inch to thirteen inches in diameter. Adjustable gate valve lockouts can be adjusted to fit any handle ranging from one inch to six-plus inches in diameter.
- *Ball valve lockouts.* There are several different configurations available to lock ball valves, all of which are designed to prevent rotation of the valve handle. The three

major configurations available are a wedge shape for one-inch to three-inch valves, a lockout that completely covers three-eighths-inch to eight-inch ball valve handles, and a universal lockout that can be applied to quarter-turn valves of varying sizes and geometric handle dimensions. All three types of ball valve lockouts can be installed by sliding the lockout device over the ball valve handle and securing it with a padlock.

- *Butterfly valve lockouts.* The butterfly valve lockout functions in a similar manner to the ball valve lockout. The polypropylene lockout device is placed over the valve handle and secured with a padlock. This type of lockout has been commonly used in the bottling industry.

A major difference between valve-specific lockout devices and the padlocked chain or cable lockouts discussed earlier is that valve-specific lockouts do not need to be secured to an anchoring device in the floor or the piping system. In addition, valve-specific lockouts eliminate potential tripping or access hazards that may be caused by chains or cable lockouts applied to valves located near walkways or frequently maintained equipment.

Valve-specific lockout devices are available in a variety of colors, which can be useful in distinguishing different valves. For example, different-colored lockouts can be used to distinguish the type of liquid passing through the valve (e.g., treated, untreated, potable, chemical) or to identify the party responsible for maintaining the lockout. Implementing a system of different-colored locks on operating valves can increase system security by reducing the likelihood of an operator inadvertently opening the wrong valve and causing a problem in the system.

Security for Vents

Vents are installed in some aboveground chemical storage areas to allow safe venting of off-gases. The specific vent design for any given application will vary depending on the design of the chemical storage vessel. However, every vent consists of an open air connection between the storage container and the outside environment. Although these air exchange vents are an integral part of covered or underground chemical storage containers, they also represent a potential security threat. Improving vent security by making the vents tamper resistant or by adding other security features, such as security screens or security covers, can enhance the security of the entire chemical processing system.

Many municipalities already have specifications for vent security at their local chemical industrial assets. These specifications typically include the following requirements:

- Vent openings are to be angled down or shielded to minimize the entrance of surface and/or rainwater into the vent through the opening.
- Vent designs are to include features to exclude insects, birds, animals, and dust.
- Corrosion-resistant materials are to be used to construct the vents.

Visual Surveillance Monitoring

Visual surveillance is used to detect threats through continuous observation of important or vulnerable areas of an asset. The observations can also be recorded for later review or use (for example, in court proceedings). Visual surveillance systems can be used to monitor various parts of collection, distribution, or treatment systems, including the perimeter of a facility, outlying pumping stations, or entry or access points into specific buildings. These systems are also useful in recording individuals who enter or leave a facility, thereby helping to identify unauthorized access. Images can be transmitted live to a monitoring station, where they can be monitored in real time, or they can be recorded and reviewed later. Many chemical facilities have found that a combination of electronic surveillance and security guards provides an effective means of facility security.

Visual surveillance is provided through a closed circuit television (CCTV) system, in which the capture, transmission, and reception of an image is localized within a closed "circuit." This is different than other broadcast images, such as over-the-air television, which is broadcast over the air to any receiver within range.

At a minimum, a CCTV system consists of

- one or more cameras
- a monitor for viewing the images
- a system for transmitting the images from the camera to the monitor

Specific attributes and features of camera systems, lenses, and lighting systems are presented in table 9.11.

CHEMICAL AND CHEMICAL WASTE MONITORING DEVICES

Earlier it was pointed out that proper security preparation really comes down to a three-legged approach: delay, respond, and detect. The third leg of security, to detect, is discussed in this section.

Sensors for Monitoring Chemical and Radiological Contamination

Toxicity tests have traditionally been used to monitor chemical wastewater effluent streams for National Pollutant Discharge Elimination System (NPDES) permit compliance. This procedure usually includes pretreatment requirements. That is, the

Table 9.11. Attributes of cameras, lenses, and lighting systems

Attribute	*Discussion*
Camera Systems	
Camera type	Major factors in choosing the correct camera are the resolution of the image required and lighting of the area to be viewed. • Solid state (including charge coupled devices, charge priming device, charge injection device, and metal oxide substrate). These cameras are becoming predominant in the marketplace because of their high resolution and their elimination of problems inherent in tube cameras. • Thermal. These cameras are designed for night vision. They require no light and use differences in temperature between objects in the field of view to produce a video image. Resolution is low compared to other cameras, and the technology is currently expensive relative to other technologies. • Tube. These cameras can provide high resolution but burn out and must be replaced after 1–2 years. In addition, tube performance can degrade over time. Finally, tube cameras are prone to burn images in the tube, requiring replacement.
Resolution (the ability to see fine details)	User must determine the amount of resolution required depending on the level of detail required for threat determination. A high definition focus with a wide field of vision will give an optimal viewing area.
Field of vision width	Cameras are designed to cover a defined field of vision, which is usually defined in degrees. The wider the field of vision, the more area a camera will be able to monitor.
Type of image produced (color, black and white, thermal)	Color images may allow the identification of distinctive markings, while black-and-white images may provide sharper contrast. Thermal imaging allows the identification of heat sources (such as human beings or other living creatures) from low-light environments; however, thermal images are not effective in identifying specific individuals (i.e., for subsequent legal processes).
Pan/tilt/zoom (PTZ)	Panning (moving the camera in a horizontal plane), tilting (moving the camera in a vertical plane), and zooming (moving the lens to focus on objects that are at different distances from the camera) allow the camera to follow a moving object. Different systems allow these functions to be controlled manually or automatically. Factors to be considered in PTZ cameras are the degree of coverage for pan and tilt functions and the power of the zoom lens.
Lenses	
Format	Lens format determines the maximum image size to be transmitted.
Focal length	This is the distance from the lens to the center of the focus. The greater the focal length, the higher the magnification, but the narrower the field of vision.
F number	F number is the ability to gather light. Smaller F numbers may be required for outdoor applications where light cannot be controlled as easily.
Distance and width approximation	The distance and width approximations are used to determine the geometry of the space that can be monitored at the best resolution.

Attribute	*Discussion*
Lighting Systems	
Intensity	Light intensity must be great enough for the camera type to produce sharp images. Light can be generated from natural or artificial sources. Artificial sources can be controlled to produce the amount and distribution of light required for a given camera and lens.
Evenness	Light must be distributed evenly over the field of view so that there are no darker or shadowy areas. If there are lighter vs. darker areas, brighter areas may appear washed out (i.e., details cannot be distinguished), while no specific objects can be viewed from darker areas.
Location	Light sources must be located above the camera so that light does not shine directly into the camera.

Source: USEPA (2005).

chemical facility must treat its waste stream to a certain level of safety to ensure that downstream wastewater treatment plant processes are not affected by industrial waste products.

Chemical sensors that can be used to identify potential threats to process water and industrial wastewater systems include inorganic monitors (e.g., chlorine analyzer), organic monitors (e.g., total organic carbon analyzer), and toxicity meters. Radiological meters can be used to measure concentrations of several different radioactive species.

Monitoring can be conducted using either portable or fixed-location sensors. Fixed-location sensors are usually used as part of a continuous, on-line monitoring system. Continuous monitoring has the advantage of enabling immediate notification when there is an upset. However, the sampling points are fixed and only certain points in the system can be monitored. In addition, the number of monitoring locations needed to capture the physical, chemical, and biological complexity of a system can be prohibitive. The use of portable sensors can overcome this problem of monitoring many points in the system. Portable sensors can be used to analyze grab samples at any point in the system, but have the disadvantage that they provide measurements only at one point in time.

Radiation Detection Equipment

Radioactive substances (radionuclides) are known health hazards that emit energetic waves and/or particles that can cause both carcinogenic and noncarcinogenic health effects. Radionuclides pose unique threats to source water supplies and chemical processing, storage, or distribution systems because radiation emitted from radionuclides in chemical or industrial waste systems can affect individuals through several pathways: by direct contact with, ingestion or inhalation of, or external exposure to, the contaminated waste stream. While radiation can occur naturally in some cases due to the decay of some minerals, intentional and nonintentional releases of man-made radionuclides into chemical water feed or chemical processing streams is also a realistic threat.

Threats to chemical facilities from radioactive contamination could involve two major scenarios. First, the facility or its assets could be contaminated, preventing workers from accessing and operating the facility/assets. Second, the feed water supply could be contaminated. These two scenarios require different threat reduction strategies. The first scenario requires that facilities monitor for radioactive substances as they are brought on-site; the second requires that feed water assets be monitored for radioactive contamination. While the effects of radioactive contamination are basically the same under both threat types, each of these threats requires different types of radiation monitoring and different types of equipment.

Radiation Detection Equipment for Monitoring Water Assets

Most water systems are required to monitor for radioactivity and certain radionuclides, and to meet maximum contaminant levels (MCLs) for these contaminants, to comply with the Safe Drinking Water Act (SDWA). Currently, USEPA requires drinking water to meet MCLs for beta/photon emitters (includes gamma radiation), alpha particles, combined radium 226/228, and uranium. However, this monitoring is required only at entry points into the system. In addition, after the initial sampling requirements, only one sample is required every three to nine years, depending on the contaminant type and the initial concentrations.

While this is adequate to monitor for long-term protection from overall radioactivity and specific radionuclides in drinking water, it may not be adequate to identify short-term spikes in radioactivity, such as from spills, accidents, or intentional releases. In addition, compliance with the SDWA requires analyzing water samples in a laboratory, which results in a delay in receiving results. In contrast, security monitoring is more effective when results can be obtained quickly in the field. In addition, monitoring for security purposes does not necessarily require that the specific radionuclides causing the contamination be identified. Thus, for security purposes, it may be more appropriate to monitor for non-radionuclide-specific radiation using either portable field meters, which can be used as necessary to evaluate grab samples, or on-line systems, which can provide continuous monitoring of a system.

Ideally, measuring radioactivity in water assets in the field would involve minimal sampling and sample preparation. However, the physical properties of specific types of radiation combined with the physical properties of water make evaluating radioactivity in water assets in the field somewhat difficult. For example, alpha particles can only travel short distances and they cannot penetrate through most physical objects. Therefore, instruments designed to evaluate alpha emissions must be specially designed to capture emissions at a short distance from the source, and they must not block alpha emissions from entering the detector. Gamma radiation does not have the same types of physical properties, and thus it can be measured using different detectors.

Measuring different types of radiation is further complicated by the relationship between the radiation's intrinsic properties and the medium in which the radiation is being measured. For example, gas-flow proportional counters are typically used to evaluate gross alpha and beta radiation from smooth, solid surfaces, but due to the fact that water is not a smooth surface, and because alpha and beta emissions are relatively short range and can be attenuated within the water, these types of counters are not appropriate for measuring alpha and beta activity in water. An appropriate method for measuring alpha and beta radiation in water is by using a liquid scintillation counter. However, this requires mixing an aliquot of water with a liquid scintillation "cocktail." The liquid scintillation counter is a large, sensitive piece of equipment, so it is not appropriate for field use. Therefore, measurements for alpha and beta radiation from water assets are not typically made in the field.

Unlike the problems associated with measuring alpha and beta activity in water in the field, the properties of gamma radiation allow it to be measured relatively well in water samples in the field. The standard instrumentation used to measure gamma radiation from water samples in the field is a sodium iodide (NaI) scintillator.

Although the devices outlined above are the most commonly used for evaluating total alpha, beta, and gamma radiation, other methods and other devices can be used. In addition, local conditions (i.e., temperature, humidity) or the properties of the specific radionuclides emitting the radiation may make other types of devices or other methods more optimal to achieve the goals of the survey than the devices noted above. There, experts or individual vendors should be consulted to determine the appropriate measurement device for any specific application.

An additional factor to consider when developing a program to monitor for radioactive contamination in feed water assets is whether to take regular grab samples or sample continuously. For example, portable sensors can be used to analyze grab samples at any point in the system, but have the disadvantage that they provide measurements only at one point in time. On the other hand, fixed-location sensors are usually used as part of a continuous, on-line monitoring system. These systems continuously monitor a feed or process water asset, and could be outfitted with some type of alarm system that would alert operators if radiation increased above a certain threshold. However, the sampling points are fixed and only certain points in the system can be monitored. In addition, the number of monitoring locations needed to capture the physical and radioactive complexity of a system can be prohibitive.

COMMUNICATION INTEGRATION DEVICES

In this section, those devices necessary for communication and integration of industrial chemical processing operations, such as electronic controllers, two-way radios, and wireless data communications are discussed. Typically, SCADA systems would also be discussed in this section; however, SCADA was discussed in detail in chapter 7.

In regard to security applications, electronic controllers are used to automatically activate equipment (such as lights, surveillance cameras, audible alarms, or locks) when they are triggered. Triggering could be in response to a variety of scenarios, including tripping of an alarm or a motion sensor, breaking of a window or a glass door, variation in vibration sensor readings, or simply through input from a timer.

Two-way wireless radios allow two or more users that have their radios tuned to the same frequency to communicate instantaneously with each other without the radios being physically linked together with wires or cables.

Wireless data communications devices are used to enable transmission of data between computer systems and/or between a SCADA server and its sensing devices, without individual components being physically linked together via wires or cables. In industrial chemical processing systems, these devices are often used to link remote monitoring stations (i.e., SCADA components) or portable computers (i.e., laptops) to computer networks without using physical wiring connections.

Electronic Controllers

An electronic controller is a piece of electronic equipment that receives incoming electric signals and uses preprogrammed logic to generate electronic output signals based on the incoming signals. While electronic controllers can be implemented for any application that involves inputs and outputs (for example, control of a piece of machinery in a factory), in a security application, these controllers essentially act as the system's "brain" and can respond to specific security-related inputs with preprogrammed output response. These systems combine the control of electronic circuitry with a logic function such that circuits are opened and closed (and thus equipment is turned on and off) through some preprogrammed logic. The basic principle behind the operation of an electrical controller is that it receives electronic inputs from sensors or any device generating an electrical signal (for example, electrical signals from motion sensors), and then uses its preprogrammed logic to produce electrical outputs (for example, these outputs could turn on power to a surveillance camera or to an audible alarm). Thus, these systems automatically generate a preprogrammed, logical response to a preprogrammed input scenario.

The three major types of electronic controllers are timers, electromechanical relays, and programmable logic controllers (PLCs), which are often called "digital relays." Each of these types of controller is discussed in more detail below.

Timers use internal signal/inputs (in contrast to externally generated inputs) and generate electronic output signals at certain times. More specifically, timers control electric current flow to any application to which they are connected, and can turn the current on or off on a schedule pre-specified by the user. Typical timer range (the amount of time that can be programmed to elapse before the timer activates linked equipment) is from 0.2 seconds to ten hours, although some of the more advanced timers have ranges of up to sixty hours. Timers are useful in fixed applications that don't require frequent schedule changes. For example, a timer can be used to turn on the lights in a room or building at a certain time every day. Timers are usually connected to their own power supply (usually 120–240 V).

In contrast to timers, which have internal triggers based on a regular schedule, electromechanical relays and PLCs have both external inputs and external outputs. However, PLCs are more flexible and more powerful than are electromechanical relays, and thus this section focuses primarily on PLCs as the predominant technology for security-related electronic control applications.

Electromechanical relays are simple devices that use a magnetic field to control a switch. Voltage applied to the relay's input coil creates a magnetic field, which attracts an internal metal switch. This causes the relay's contacts to touch, closing the switch and completing the electrical circuit. This activates any linked equipment. These types of systems are often used for high-voltage applications, such as in some automotive and other manufacturing processes.

Two-Way Radios

Two-way radios, as discussed here, are limited to a direct unit-to-unit radio communication, either via single unit-to-unit transmission and reception, or via multiple handheld units to a base station radio contact and distribution system. Radio frequency spectrum limitations apply to all handheld units, and are directed by the FCC. This also distinguishes a handheld unit from a base station or base station unit (such as those used by an amateur [ham] radio operator), which operates under different wave-length parameters.

Two-way radios allow a user to contact another user or group of users instantly on the same frequency, and to transmit voice or data without the need for wires. They use "half-duplex" communications, or communication that can be only transmitted or received; they cannot transmit and receive simultaneously. In other words, only one person may talk, while other personnel with radio(s) can only listen. To talk, the user depresses the talk button and speaks into the radio. The audio then transmits the voice wirelessly to the receiving radios. When the speaker has finished speaking and the channel has cleared, users on any of the receiving radios can transmit, either to answer the first transmission or to begin a new conversation. In addition to carrying voice data, many types of wireless radios also allow the transmission of digital data, and these radios may be interfaced with computer networks that can use or track these data. For example, some two-way radios can send information such as global positioning system (GPS) data, or the ID of the radio. Some two-way radios can also send data through a SCADA system.

Wireless radios broadcast these voice or data communications over the airwaves from the transmitter to the receiver. While this can be an advantage in that the signal emanates in all directions and does not need a direct physical connection to be received at the receiver, it can also make the communications vulnerable to being blocked, intercepted, or otherwise altered. However, security features are available to ensure that the communications are not tampered with.

Wireless Data Communications

A wireless data communication system consists of two components: a wireless access point (WAP) and a wireless network interface card (sometimes also referred to as a "client"), which work together to complete the communications link. These wireless systems can link electronic devices, computers, and computer systems together using radio waves, thus eliminating the need for these individual components to be directly connected together through physical wires. While wireless data communications have widespread application in industrial chemical industries, they also have limitations. First, wireless data connections are limited by the distance between components (radio waves scatter over a long distance and cannot be received efficiently, unless special directional antennae are used). Second, these devices only function if the individual components are in a direct line of sight with each other, since radio waves are affected by interference from physical obstructions. However, in some cases, repeater units can be used to amplify and retransmit wireless signals to circumvent these problems. The two components of wireless devices are discussed in more detail below.

1. *WAP.* The WAP provides the wireless data communication service. It usually consists of a housing (which is constructed from plastic or metal depending on the environment it will be used in) containing a circuit board, flash memory that holds software, one of two external ports to connect to existing wired networks, a wireless radio transmitter/receiver, and one or more antenna connections. Typically, the WAP requires a one-time user configuration to allow the device to interact with the local area network (LAN). This configuration is usually done via a Web-driven software application which is accessed via a computer.
2. *Wireless network interface card/client.* A wireless card is a piece of hardware that is plugged in to a computer and enables that computer to make a wireless network connection. The card consists of a transmitter, functional circuitry, and a receiver

for the wireless signal, all of which work together to enable communication between the computer, its wireless transmitter/receiver, and its antenna connection. Wireless cards are installed in a computer through a variety of connections, including USB adapters, or laptop cardbus (PCMCIA) or desktop peripheral (PCI) cards. As with the WAP, software is loaded onto the user's computer, allowing configuration of the card so that it may operate over the wireless network.

Two of the primary applications for wireless data communications systems are to enable mobile or remote connections to a LAN, and to establish wireless communications links between SCADA remote telemetry units (RTUs) and sensors in the field. Wireless card connections are usually used for LAN access from mobile computers. Wireless cards can also be incorporated into RTUs to allow them to communicate with sensing devices that are located remotely.

CYBER PROTECTION DEVICES

Various cyber protection devices are currently available for use in protecting utility computer systems. These protection devices include anti-virus and pest eradication software, firewalls, and network intrusion hardware/software. These products are discussed in this section.

Anti-Virus and Pest Eradication Software

Anti-virus programs are designed to detect, delay, and respond to programs or pieces of code that are specifically designed to harm computers. These programs are known as "malware." Malware can include computer viruses, worms, and Trojan horse programs (programs that appear to be benign but have hidden harmful effects).

Pest eradication tools are designed to detect, delay, and respond to "spyware" (strategies that websites use to track user behavior, such as by sending "cookies" to the user's computer) and hacker tools that track keystrokes (keystroke loggers) or passwords (password crackers).

Viruses and pests can enter a computer system through the Internet or through infected floppy discs or CDs. They can also be placed onto a system by insiders. Some of these programs, such as viruses and worms, can then move within a computer's drives and files, or between computers if the computers are networked to each other. This malware can deliberately damage files, utilize memory and network capacity, crash application programs, and initiate transmissions of sensitive information from a PC. While the specific mechanisms of these programs differ, they can infect files and even the basic operating program of the computer firmware/hardware.

The most important features of an anti-virus program are its abilities to identify potential malware and to alert a user before infection occurs, as well as its ability to respond to a virus already resident on a system. Most of these programs provide a log so that the user can see what viruses have been detected and where they were detected. After detecting a virus, the anti-virus software may delete the virus automatically, or it may prompt the user to delete the virus. Some programs will also fix files or programs damaged by the virus.

Various sources of information are available to inform the general public and computer system operators about new viruses being detected. Since anti-virus programs use signatures (or snippets of code or data) to detect the presence of a virus, periodic updates are required to identify new threats. Many anti-virus software providers offer free upgrades that are able to detect and respond to the latest viruses.

Firewalls

A firewall is an electronic barrier designed to keep computer hackers, intruders, or insiders from accessing specific data files and information on a facility's computer network or other electronic/computer systems. Firewalls operate by evaluating and then filtering information coming through a public network (such as the Internet) into the facility's computer or other electronic system. This evaluation can include identifying the source or destination addresses and ports, and allowing or denying access based on this identification.

There are two methods used by firewalls to limit access to the facility's computers or other electronic systems from the public network:

- The firewall may deny all traffic unless it meets certain criteria.
- The firewall may allow all traffic through unless it meets certain criteria.

A simple example of the first method is to screen requests to ensure that they come from an acceptable (i.e., previously identified) domain name and Internet protocol address. Firewalls may also use more complex rules that analyze the application data to determine if the traffic should be allowed through. For example, the firewall may require user authentication (i.e., use of a password) to access the system. How a firewall determines what traffic to let through depends on which network layer it operates at and how it is configured. Some of the pros and cons of various methods to control traffic flowing in and out of the network are provided in table 9.12.

Firewalls may be a piece of hardware, a software program, or an appliance card that contains both. Advanced features that can be incorporated into firewalls allow for the tracking of attempts to log on to the local area network system. For example, a report of successful and unsuccessful log-on attempts may be generated for the computer specialist to analyze. For systems with mobile users, firewalls allow remote access to the private network by the use of secure log-on procedures

Table 9.12. Pros and cons of various firewall methods for controlling network access

Method	*Description*	*Pros*	*Cons*
Packet filtering	Incoming and outgoing packets (small chunks of data) are analyzed against a set of filters. Packets that make it through the filters are sent to the requesting system and all others are discarded. There are two type of packet filtering: static (more common) and dynamic.	Static filtering is relatively inexpensive, and little maintenance required. It is well suited for closed environments where access to or from multiple addresses is not allowed.	Leaves permanent open holes in the network; allows direct connection to internal hosts by external sources; offers no user authentication; method can be unmanageable in large networks.
Proxy service	Information from the Internet is retrieved by the firewall and then sent to the requesting system and vice versa. In this way, the firewall can limit the information made known to the requesting system, making vulnerabilities less apparent.	Only allows temporary open holes in the network perimeter. Can be used for all types of internal protocol services.	Allows direct connections to internal hosts by external clients; offers no user authentication
Stateful pattern recognition	This method examines and compares the contents of certain key parts of an information packet against a database of acceptable information. Information traveling from inside the firewall to the outside is monitored for specific defining characteristics, then incoming information is compared to these characteristics. If the comparison yields a reasonable match, the information is allowed through. If not, the information is discarded.	Provides a limited time window to allow pockets of information to be sent; does not allow any direct connections between internal and external hosts; supports user-level authentication.	Slower than packet filtering; does not support all types of connections.

Source: USEPA (2005).

and authentication certificates. Most firewalls have a graphical user interface for managing the firewall.

In addition, new Ethernet firewall cards that fit in the slot of an individual computer bundle additional layers of defense (like encryption and permit/deny) for individual computer transmissions to the network interface function. These new cards have only a slightly higher cost than traditional network interface cards.

Network Intrusion Hardware/Software

Network intrusion detection and prevention systems are software- and hardware-based programs designed to detect unauthorized attacks on a computer network system.

While other applications, such as firewalls and anti-virus software, share similar objectives with network intrusion systems, network intrusion systems provide a deeper layer of protection beyond the capabilities of these other systems because they evaluate patterns of computer activity rather than specific files.

It is worth noting that attacks may come from either outside or within the system (i.e., from an insider), and that network intrusion detection systems may be more applicable for detecting patterns of suspicious activity from inside a facility (i.e., accessing sensitive data, etc.) than are other information technology solutions.

Network intrusion detection systems employ a variety of mechanisms to evaluate potential threats. The types of search and detection mechanisms are dependent upon the level of sophistication of the system. Some of the available detection methods include the following:

- *Protocol analysis.* Protocol analysis is the process of capturing, decoding, and interpreting electronic traffic. The protocol analysis method of network intrusion detection involves the analysis of data captured during transactions between two or more systems or devices, and the evaluation of these data to identify unusual activity and potential problems. Once a problem is isolated and recorded, problems or potential threats can be linked to pieces of hardware or software. Sophisticated protocol analysis will also provide statistics and trend information on the captured traffic.
- *Traffic anomaly detection.* Traffic anomaly detection identifies potential threatening activity by comparing incoming traffic to "normal" traffic patterns and identifying deviations. It does this by comparing user characteristics against thresholds and triggers defined by the network administrator. This method is designed to detect attacks that span a number of connections, rather than a single session.
- *Network honeypot.* This method establishes nonexistent services in order to identify potential hackers. A network honeypot impersonates services that don't exist by sending fake information to people scanning the network. It identifies the attackers

when they attempt to connect to the service. There is no reason for legitimate traffic to access these resources because they don't exist; therefore any attempt to access them constitutes an attack.

- *Anti-intrusion detection system evasion techniques.* These methods are designed for attackers who may be trying to evade intrusion detection system scanning. They include methods called IP defragmentation, TCP streams reassembly, and deobfuscation.

While these detection systems are automated, they can only indicate patterns of activity, and a computer administrator or other experienced individual must interpret activities to determine whether or not they are potentially harmful. Monitoring the logs generated by these systems can be time consuming, and there may be a learning curve to determine a baseline of "normal" traffic patterns from which to distinguish potential suspicious activity.

REFERENCES AND RECOMMENDED READING

Garcia, M. L. 2001. *The design and evaluation of physical protection systems.* Butterworth-Heinemann.

International Bottled Water Association (IBWA). 2004. *Bottled water safety and security.* Alexandria, VA: International Bottled Water Association.

U.S. Environmental Protection Agency (USEPA). 2005. *Water and wastewater security product guide.* www.epa.gov/safewater/security/guide (accessed June 2006).

10

The Paradigm Shift

The 9/11 shift: There is a new worldview in the making.

The events of 9/11 dramatically changed this nation and focused us on combating terrorism. As a result, in 2003 and subsequent years, the Department of Homeland Security (DHS) in conjunction with members from the general public, state, and local agencies, and private groups concerned with the safety of critical infrastructures, established a Water Security Working Group (WSWG) under the U.S. Environmental Protection Agency (USEPA) to consider and make recommendations on infrastructure security issues. For example, the WSWG identified active and effective security practices for critical infrastructure, and provided an approach for adopting these practices. It also recommended mechanisms to provide incentives that facilitate broad and receptive response among critical infrastructure sectors to implement active and effective security practices. Finally, WSWG recommended mechanisms to measure progress and achievements in implementing active and effective security practices, and to identify barriers to implementation.

The WSWG recommendations on security are structured to maximize benefits to critical industries by emphasizing actions that have the potential both to improve the quality or reliability of service and to enhance security. These recommendations, based on original recommendations from the 2003 National Drinking Water Advisory Council (NDWAC), were designed primarily, as the name suggests, for use by water systems of all types and sizes, including systems that serve less than 3,300 people. However, it is the authors' opinion, based on personal experience, that NDWAC's recommendations, when properly adapted to applicable circumstances, can be applied to any and all critical infrastructure sectors, including the chemical industry.

The NDWAC identified fourteen features of active and effective security programs that are important to increasing security and relevant across the broad range of utility circumstances and operating conditions. USEPA (2003) points out that the fourteen features are, in many cases, consistent with the steps needed to maintain technical, management, and operational performance capacity related to overall water quality; these steps can be applied to other critical infrastructures as well. Many facilities may be able to adopt some of the features with minimal, if any, capital investment.

FOURTEEN FEATURES OF ACTIVE AND EFFECTIVE SECURITY

It is important to point out that the fourteen features of active and effective programs emphasize that "one size does not fit all" and that there will be variability in security approaches and tactics among chemical facilities, based on industry-specific circumstances and operating conditions. These features

- are sufficiently flexible to apply to all chemical industries, regardless of size
- incorporate the idea that active and effective security programs should have measurable goals and timelines
- allow flexibility for chemical industrial facilities to develop specific security approaches and tactics that are appropriate to industry-specific circumstances

Chemical facilities can differ in many ways, including the following:

- transportation supply source (rail, air, water, or ground)
- number of supply sources
- chemical processing capacity
- operation risk
- location risk
- security budget
- spending priorities
- political and public support
- legal barriers
- public versus private ownership

Chemical industrial facilities should address security in an informed and systematic way, regardless of these differences. Chemical facilities need to fully understand the specific local circumstances and conditions under which they operate, and to develop a security program tailored to those conditions. The goal in identifying common features of active and effective security programs is to achieve consistency in security program outcomes among chemical industrial facilities, while allowing for and encouraging facilities to develop utility-specific security approaches and tactics. The features are based on a comprehensive "security management layering system" approach that incorporates a combination of public involvement and awareness; partnerships; and physical, chemical, operational, and design controls to increase overall program performance. They address industry security in four functional categories: *organization, operation, infrastructure,* and *external.* These functional categories are discussed in greater detail below.

- *Organizational.* There is always something that can be done to improve security. Even when resources are limited, the simple act of increasing organizational attentiveness to security may reduce vulnerability and increase responsiveness. Preparedness itself can help deter attacks. The first step to achieving preparedness is to make security a part of the organizational culture, so that it is in the day-to-day thinking of frontline employees, emergency responders, and management of every chemical facility in this country. To successfully incorporate security into "business as usual," there must be a strong commitment to security by organization leadership and by the supervising body, such as the board of stockholders. The following features address how a security culture can be incorporated into an organization.
- *Operational.* In addition to having a strong culture and awareness of security within an organization, an active and effective security program makes security part of operational activities, from daily operations (such as monitoring physical access controls) to scheduled annual reassessments. Chemical industries will often find that by implementing security into operations they can also reap cost benefits and improve the quality or reliability of the chemical service.
- *Infrastructure.* These recommendations advise utilities to address security in all elements of chemical industry infrastructure—from source chemical to transportation and through processing and product delivery.
- *External.* Strong relationships with response partners and the public strengthen security and public confidence. Two of the recommended features of active and effective security programs address this need.

Feature 1. Make an explicit and visible commitment of the senior leadership to security.

Chemical industrial facilities should create an explicit, easily communicated, enterprise-wide commitment to security, which can be done through

- incorporating security into a utility-wide mission or vision statement addressing the full scope of an active and effective security program—that is, protection of worker/public health, worker/public safety, and public confidence—that is part of core day-to-day operations
- developing an enterprise-wide security policy, or set of policies

Chemical industries should use the process of making a commitment to security as an opportunity to raise awareness of security throughout the organization, making the commitment visible to all employees and customers, and to help every facet of the enterprise to recognize the contribution they can make to enhancing security.

Feature 2. Promote security awareness throughout the organization.

The objective of a security culture should be to make security awareness a normal, accepted, and routine part of day-to-day operations. Examples of tangible efforts include:

- conducting employee training
- incorporating security into job descriptions
- establishing performance standards and evaluations for security
- creating and maintaining a security tip line and suggestion box for employees
- making security a routine part of staff meetings and organization planning
- creating a security policy

Feature 3. Assess vulnerabilities and periodically review and update vulnerability assessments to reflect changes in potential threats and vulnerabilities.

Because circumstances change, chemical industrial facilities should maintain their understanding and assessment of vulnerabilities as a "living document," and continually adjust their security enhancement and maintenance priorities. Chemical industrial facilities should consider their individual circumstances and establish and implement a schedule for review of their vulnerabilities.

Assessments should take place once every three to five years at a minimum. Chemical industries may be well served by doing assessments annually. The basic elements of sound vulnerability assessments are

- characterization of the chemical processing system, including its mission and objectives
- identification and prioritization of adverse consequences to avoid
- determination of critical assets that might be subject to malevolent acts that could result in undesired consequences
- assessment of the likelihood (qualitative probability) of such malevolent acts from adversaries
- evaluation of existing countermeasures
- analysis of current risk and development of a prioritized plan for risk reduction

Feature 4. Identify security priorities and, on an annual basis, identify the resources dedicated to security programs and planned security improvements, if any.

Dedicated resources are important to ensure a sustained focus on security. Investment in security should be reasonable considering utilities' specific circumstances. In some circumstances, investment may be as simple as increasing the amount of time and attention that executives and managers give to security. Where threat potential or potential consequences are greater, greater investment likely is warranted.

This feature establishes the expectation that chemical industrial facilities should, through their annual capital, operations, maintenance, and staff resources plans, identify and set aside resources consistent with their specific identified security needs. Security priorities should be clearly documented and should be reviewed with utility executives at least once per year as part of the traditional budgeting process.

Feature 5. Identify managers and employees who are responsible for security and establish security expectations for all staff.

- Explicit identification of security responsibilities is important for development of a security culture with accountability.
- At minimum, chemical industrial facilities should identify a single, designated individual responsible for overall security, even if other security roles and responsibilities will likely be dispersed throughout the organization.
- The number and depth of security-related roles will depend on a facility's specific circumstances.

Feature 6. Establish physical and procedural controls to restrict access to chemical industrial infrastructure to only those conducting authorized, official business and to detect unauthorized physical intrusions.

Examples of physical access controls include fencing critical areas, locking gates and doors, and installing barriers at site access points. Monitoring for physical intrusion can include maintaining well-lighted facility perimeters, installing motion detectors, and utilizing intrusion alarms. The use of neighborhood watches, regular employee rounds, and arrangements with local police and fire departments can support identifying unusual activity in the vicinity of facilities.

Examples of procedural access controls include inventorying keys, changing access codes regularly, and requiring security passes to pass gates and access sensitive areas. In addition, facilities should establish the means to readily identify all employees, including contractors and temporary workers with unescorted access to facilities.

Feature 7. Create employee protocols for detection of contamination consistent with the recognized limitations in current contaminant detection, monitoring, and surveillance technology.

Until progress can be made in development of practical and affordable online contaminant monitoring and surveillance systems, most chemical industrial facilities must use other approaches to contaminant monitoring and surveillance. This includes monitoring data of physical and chemical contamination surrogates, pressure change abnormalities, free and total chlorine residual, temperature, dissolved oxygen, and conductivity.

Many facilities already measure the above parameters (and many others) on a regular basis to control plant operations and confirm chemical mixture quality; more closely monitoring these parameters may create operational benefits for facilities that extend far beyond security, such as reducing operating costs and chemical usage. Chemical industrial facilities also should thoughtfully monitor customer complaints and improve connections with local public health networks to detect public health anomalies. Customer complaints and public health anomalies are important ways to detect potential contamination problems and other environmental quality concerns.

Feature 8. Define security-sensitive information; establish physical, electronic, and procedural controls to restrict access to security-sensitive information; detect unauthorized access; and ensure information and communications systems will function during emergency response and recover.

Protecting IT systems largely involves using physical hardening and procedural steps to limit the number of individuals with authorized access and to prevent access by unauthorized individuals. Examples of physical steps to harden SCADA and IT networks include installing and maintaining firewalls, and screening the network for viruses. Examples of procedural steps include restricting remote access to data networks, and safeguarding critical data through backups and storage in safe places. Facilities should strive for continuous operation of IT and telecommunications systems, even in the event of an attack, by providing uninterruptible power supply and backup systems, such as satellite phones.

In addition to protecting IT systems, security-sensitive information should be identified and restricted to the appropriate personnel. Security-sensitive information could be contained within

- facility maps and blueprints
- operations details
- hazardous material utilization and storage
- tactical level security program details

- any other information on utility operations or technical details that could aid in planning or executing an attack

Those identifying security-sensitive information should consider all ways that facilities might use and make public information (e.g., many chemical industrial facilities may at times engage in competitive bidding processes for construction of new facilities or infrastructure). Finally, information critical to the continuity of day-to-day operations should be identified and backed up.

Feature 9. Incorporate security considerations into decisions about acquisition, repair, major maintenance, and replacement of physical infrastructure; include consideration of opportunities to reduce risk through physical hardening and adoption of inherently lower-risk design and technology options.

Prevention is a key aspect of enhancing security. Consequently, consideration of security issues should begin as early as possible in facility construction (i.e., it should be a factor in building plans and designs). However, to incorporate security considerations into design choices, chemical facilities need information about the types of security design approaches and equipment that are available and the performance of these designs and equipment in multiple dimensions. For example, chemical facilities would want to evaluate not just the way that a particular design might contribute to security, but would also look at how that design would affect the efficiency of day-to-day plant operations and worker safety.

Feature 10. Monitor available threat-level information and escalate security procedures in response to relevant threats.

Monitoring threat information should be a regular part of a security program manager's job, and utility-, facility-, and region-specific threat levels and information should be shared with those responsible for security. As part of security planning, chemical facilities should develop systems to access threat information and procedures that will be followed in the event of increased industry or facility threat levels, and should be prepared to put these procedures in place immediately, so that adjustments are seamless. Involving local law enforcement and FBI is critical.

Chemical facilities should investigate what networks and information sources might be available to them locally, and at the state and regional level. If a facility cannot gain access to some information networks, attempts should be made to align with those who can and will provide effective information to the chemical facility on a timely basis.

Feature 11. Incorporate security considerations into emergency response and recovery plans, test and review plans regularly, and update plans to reflect changes in potential threats, physical infrastructure, chemical processing operations, critical interdependencies, and response protocols in partner organizations.

Chemical facilities should maintain response and recovery plans as "living documents." In incorporating security considerations into their emergency response and recovery plans, chemical facilities also should be aware of the National Incident Management System (MIMS) guidelines, established by DHS, and of regional and local incident management commands and systems, which tend to flow from the national guidelines.

Chemical facilities should consider their individual circumstances and establish, develop, and implement a schedule for review of emergency response and recovery plans. Chemical facility plans should be thoroughly coordinated with emergency response and recovery planning in the larger community. As part of this coordination, a mutual aid program should be established to arrange in advance for exchanging resources (personnel or physical assets) among chemical industrial facilities within a region, in the event of an emergency or disaster that disrupts operation. Typically, the exchange of resources is based on a written formal mutual-aid agreement. For example, Florida's Water-Wastewater Agency Response Network (FlaWARN), deployed after Hurricane Katrina, allowed the new "utilities helping utilities" network to respond to urgent requests from Mississippi for help to bring facilities back online after the hurricane.

The emergency response and recovery plans should be reviewed and updated as needed annually. This feature also establishes the expectation that chemical facilities should test or exercise their emergency response and recovery plans regularly.

Feature 12. Develop and implement strategies for regular, ongoing security-related communications with employees, response organizations, rate-setting organizations, and customers.

An active and effective security program should address protection of public health, public safety (including infrastructure), and public confidence. Chemical facilities should create an awareness of security and an understanding of the rationale for their overall security management approach in the communities they reside in and/or serve.

Effective communication strategies consider key messages; who is best equipped/trusted to deliver the key messages; the need for message consistency, particularly during an emergency; and the best mechanisms for delivering messages and for receiving information and feedback from key partners. The key audiences for communication strategies are utility employees, response organizations, and customers.

Feature 13. Forge reliable and collaborative partnerships with the communities served, managers of critical interdependent infrastructure, response organizations, and local utilities.

Effective partnerships build collaborative working relationships and clearly define roles and responsibilities, so that people can work together seamlessly if an emergency should occur. It is important for chemical facilities within a region and neighboring regions to collaborate and establish a mutual aid program with neighboring utilities, response organizations, and sectors, such as the power sector, on which utilities rely or impact. Mutual aid agreements provide for help from other organizations that is prearranged and can be accessed quickly and efficiently in the event of a terrorist attack or natural disaster. Developing reliable and collaborative partnerships involves reaching out to managers and key staff in other organizations to build reciprocal understanding and to share information about the facility's security concerns and planning. Such efforts will maximize the efficiency and effectiveness of a mutual aid program during an emergency response effort, as the organizations will be familiar with each others' circumstances, and thus will be better able to serve each other.

It is also important for chemical facilities to develop partnerships with the communities and customers they serve. Partnerships help to build credibility within communities and establish public confidence in utility operations. People who live near chemical facility structures can be the eyes and ears of the facility, and can be encouraged to notice and report changes in operating procedures or other suspicious behaviors.

Chemical facilities and public health organizations should establish formal agreements on coordination to ensure regular exchanges of information between facilities and public health organizations, and should outline roles and responsibilities during response to and recovery from an emergency. Coordination is important at all levels of the public health community—national public health, county health agencies, and healthcare providers, such as hospitals.

Feature 14. Develop chemical facility–specific measures of security activities and achievements, and self-assess against these measures to understand and document program progress.

Although security approaches and tactics will be different depending on chemical facility–specific circumstances and operating conditions, we recommend that all chemical facilities monitor and measure a number of common types of activities and achievements, including existence of program policies and procedures, training, testing, and implementing schedules and plans.

THE FOURTEEN FEATURE MATRIX

In sidebar 10.1, a matrix of recommended measures to assess the effectiveness of a chemical facility's security program is presented. Each feature is grouped according to its functional category: organization, operation, infrastructure, and external.

Sidebar 10.1. Recommended Measures to Assess Effectiveness of Security Program

Feature	*Measure of Potential Progress*
Organizational Features	
Feature 1—Explicit commitment to security	Does a written, enterprise-wide security policy exist, and is the policy reviewed regularly and updated as needed?
Feature 2—Promote security awareness	Are incidents reported in a timely way, and are lessons learned from incident responses reviewed and, as appropriate, incorporated into future security efforts?
Feature 5—Defined security roles and employee expectations	Are managers and employees who are responsible for security identified?
Operational Features	
Feature 3—Vulnerability assessment up to date	Are reassessments of vulnerabilities made after incidents, and are lessons learned and other relevant information incorporated into security practices?
Feature 4—Security resources and implementation priorities	Are security priorities clearly identified, and to what extent do security priorities have resources assigned to them?
Feature 7—Contamination detection	Is there a protocol/procedure in place to identify and respond to suspected contamination events?
Feature 10—Threat-level-based protocols	Is there a protocol/procedure for responses that will be made if threat levels change?
Feature 11—Emergency response plan tested and up to date	Do exercises address the full range of threats—physical, cyber, and contamination—and is there a protocol/procedure to incorporate lessons learned from exercises and actual responses into updates to emergency response and recovery plans?
Feature 14—Industry-specific measures and self-assessment	Does the utility perform self-assessment at least annually?

Infrastructure Features	
Feature 6—Intrusion detection and access control	To what extent are methods to control access to sensitive assets in place?
Feature 8—Information protection and continuity	Is there a procedure to identify and control security-sensitive information? Is information correctly categorized, and how do control measures perform under testing?
Feature 9—Design and construction standards	Are security considerations incorporated into internal design and construction standards for new facilities/ infrastructure and major maintenance projects?
External Features	
Feature 12—Communications	Is there a mechanism for employees, partners, and the community to notify the facility of suspicious occurrences and other security concerns?
Feature 13—Partnerships	Have reliable and collaborative partnerships with customers, managers of independent interrelated infrastructure, and response organizations been established?

Source: USEPA (2003).

Ultimately, the goal of implementing the fourteen security features (and all other security provisions) is to create a significant improvement in the chemical industry on a national scale, by reducing vulnerabilities and therefore risks to public health from terrorist attacks and natural disasters. To create a sustainable effect, the chemical industrial sector as a whole must not only adopt and actively practice the features, but also incorporate the features into "business as usual."

REFERENCES AND RECOMMENDED READING

U.S. Environmental Protection Agency (USEPA). 2003. Active and effective water security programs. cfpub.epa.gov/safewater/watersecurity/14 features.cfm (accessed June 2006).

Index

About the Authors

Frank R. Spellman is Assistant Professor of Environmental Health at Old Dominion University. He is a professional member of the American Society of Safety Engineers, the Water Environment Federation, and the Institute of Hazardous Materials Managers. He is also a Board Certified Safety Professional and Board Certified Hazardous Materials Manager with more than 35 years of experience in environmental science and engineering. He is the author of more than fifty books, including *Water Infrastructure Protection and Homeland Security* (GI, 2007) and *Food Supply Protection and Homeland Security* (GI, 2008)

Revonna M. Bieber is currently working for the Naval Medical Center Portsmouth in the field of industrial hygiene. Her work focuses on environmental health hazards and radiologic and healthcare safety. She is the co-author, with Frank R. Spellman, of *Occupational Safety and Health* Simplified *for the Chemical Industry* (GI, 2009) and numerous other books.